An Introduction to
Numerical Methods
for the Physical Sciences

Synthesis Lectures on Engineering, Science, and Technology

Each book in the series is written by a well known expert in the field. Most titles cover subjects such as professional development, education, and study skills, as well as basic introductory undergraduate material and other topics appropriate for a broader and less technical audience. In addition, the series includes several titles written on very specific topics not covered elsewhere in the Synthesis Digital Library.

An Introduction to Numerical Methods for the Physical Sciences
Colm T. Whelan
2020

Introduction to Engineering Research
WEndy C. Crone
2020

Theory of Electromagnetic Beams
John Lekner
2020

The Search for the Absolute: How Magic Became Science
Jeffrey H. Williams
2020

The Big Picture: The Universe in Five S.T.E.P.S.
John Beaver
2020

Relativistic Classical Mechanics and Electrodynamics
Martin Land and Lawrence P. Horwitz
2019

Generating Functions in Engineering and the Applied Sciences
Rajan Chattamvelli and Ramalingam Shanmugam
2019

Transformative Teaching: A Collection of Stories of Engineering Faculty's Pedagogical Journeys
Nadia Kellam, Brooke Coley, and Audrey Boklage
2019

Ancient Hindu Science: Its Transmission and Impact on World Cultures
Alok Kumar
2019

Value Rational Engineering
Shuichi Fukuda
2018

Strategic Cost Fundamentals: for Designers, Engineers, Technologists, Estimators, Project Managers, and Financial Analysts
Robert C. Creese
2018

Concise Introduction to Cement Chemistry and Manufacturing
Tadele Assefa Aragaw
2018

Data Mining and Market Intelligence: Implications for Decision Making
Mustapha Akinkunmi
2018

Empowering Professional Teaching in Engineering: Sustaining the Scholarship of Teaching
John Heywood
2018

The Human Side of Engineering
John Heywood
2017

Geometric Programming for Design Equation Development and Cost/Profit Optimization (with illustrative case study problems and solutions), Third Edition
Robert C. Creese
2016

Engineering Principles in Everyday Life for Non-Engineers
Saeed Benjamin Niku
2016

A, B, See... in 3D: A Workbook to Improve 3-D Visualization Skills
Dan G. Dimitriu
2015

Essentials of Applied Mathematics for Scientists and Engineers
Robert G. Watts
2007

Project Management for Engineering Design
Charles Lessard and Joseph Lessard
2007

Relativistic Flight Mechanics and Space Travel
Richard F. Tinder
2006

An Introduction to Numerical Methods for the Physical Sciences
Colm T. Whelan

ISBN: 978-3-031-00957-0   paperback
ISBN: 978-3-031-02085-8   ebook
ISBN: 978-3-031-00157-4   hardcover

DOI 10.1007/978-3-031-02085-8

A Publication in the Springer series
SYNTHESIS LECTURES ON ENGINEERING, SCIENCE, AND TECHNOLOGY

Lecture #8
Series ISSN
Print 2690-0300   Electronic 2690-0327

An Introduction to
Numerical Methods
for the Physical Sciences

Colm T. Whelan
Old Dominion University

SYNTHESIS LECTURES ON ENGINEERING, SCIENCE, AND TECHNOLOGY #8

ABSTRACT

There is only a very limited number of physical systems that can be exactly described in terms of simple analytic functions. There are, however, a vast range of problems which are amenable to a computational approach. This book provides a concise, self-contained introduction to the basic numerical and analytic techniques, which form the foundations of the algorithms commonly employed to give a quantitative description of systems of genuine physical interest. The methods developed are applied to representative problems from classical and quantum physics.

KEYWORDS

differential equations, linear equations, polynomial approximations, variational principles

For my colleagues and friends:
Reiner Dreizler and James Walters

Contents

Preface

There is only a limited number of physical systems that can be exactly described in terms of simple analytic functions. There are, however, a vast range of problems that are amenable to a computational approach. This book provides a concise introduction to the essential numerical and analytic techniques which form the foundations of algorithms commonly employed to give a quantitative description of systems of genuine physical interest. Rather than providing a series of useful programming recipes the philosophy of the book is to present in a coherent way the underlying theory. I include some case studies illustrating the application to problems in classical and quantum physics.

Colm T. Whelan
Norfolk, June 2020

CHAPTER 1

Preliminaries

Before diving into the study of numerical methods and their applications, it is worthwhile to briefly think about how computers work and how we interact with them.

1.1 NUMBERS AND ERRORS

Computers store everything in bits (or binary digits). Bits have only two possible values, which are usually denoted by 0 and 1. Computers store numbers in binary as a sequence of bits, and can represent integers exactly, as long as there are enough bits but cannot store most real numbers with complete accuracy. Obviously they cannot deal with irrational numbers such as π or e but they also struggle when working with very big or very small numbers and those with large numbers of significant digits. Computers represent a floating-point real number, r, as

$$r = (-1)^s fbe,$$

where s, f, b, and e are integers; s determines the sign, f is the "*significand*" (or "*coefficient*"), b is the base (usually 2), and e is the exponent. The possible finite values that can be represented in a given format are determined by the number of digits in the significand f, the base b, and the number of digits in the exponent e. When the computer adds two floating point numbers, it first fixes them to have the same exponent then adds, clearly there is only a finite number of places of decimals that can be stored and this leads to the potential for numerical errors.

- Roundoff Errors.
  A roundoff error is the difference between the result produced by a given algorithm using exact arithmetic and the result produced by the same algorithm using finite-precision, rounded arithmetic. Such an error can grow in significance when we have a large number of repeated operations.

- Cancellation Errors.
  These occur when we subtract two almost equal numbers one from the other.

1.2 ALGORITHMS

The application of numerical methods to the description of physical problems proceeds through the formulation and implementation of algorithms, i.e., sequences of well-defined instructions, which a computer can interpret in an unambiguous way leading to a numerical solution of the problem at hand. One of the challenges you will face in translating the equations of physics into efficient and accurate computer code is that an expression that is mathematically correct may be highly susceptible to numerical errors. We have to be careful to design our algorithms to avoid such errors. I will illustrate this by two simple examples.

Example 1.1   Suppose we are looking for the roots of the quadratic

$$x^2 - 2bx + c = 0.$$

We know this has two roots

$$x_\pm = -b \pm \sqrt{b^2 - c}. \tag{1.1}$$

If $c \ll 1$, then the root x_- is very susceptible to cancellation errors. However, we may write

$$
\begin{aligned}
x_- x_+ &= b^2 - b^2 + c, \\
x_- &= \frac{c}{x_+},
\end{aligned} \tag{1.2}
$$

which gives us an expression which is much less sensitive.

Example 1.2   Suppose we want to compute the definite integral

$$I_n = \int_0^1 x^n e^x \, dx, \tag{1.3}$$

for $n = 1, \ldots, 15$. Using integration by parts it follows that

$$
\begin{aligned}
I_n &= e - n I_{n-1}, \\
I_0 &= e - 1.
\end{aligned} \tag{1.4}
$$

In Figure 1.1 we plot the result of using the recurrence relation (1.4), for increasing n, compared with the direct numerical integration of (1.3). The issue with using the recursion relation (1.4) is the "*unstable algorithm*," which magnifies the initial error at each step. If I_n is the exact value, and $\bar{I}_n$ our numerical estimate and ϵ_n the error at each step then the magnitude of the error is

$$
\begin{aligned}
|\epsilon_n| = |I_n - \bar{I}_n| &= |(e - n I_{n-1}) - |(e - n \bar{I}_{n-1})| \\
&= n |I_{n-1} - \bar{I}_{n-1}| \\
&= n |\epsilon_{n-1}| \\
&= n! |\epsilon_0|.
\end{aligned}
$$

This error becomes rapidly larger as n increases. We note from the mean value theorem for integrals that I_n will go to zero as n increases If now we rewrite (1.4)

$$I_{n-1} = \frac{1}{n} [e - I_n] \tag{1.5}$$

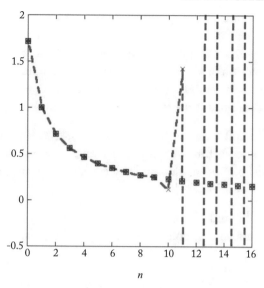

Figure 1.1: Evaluation of the integral $I_n = \int_0^1 x^n e^x dx$ using: (i) direct numerical integration, open blue squares; (ii) backward recurrence, solid red disks; and (iii) forward recurrence, green crosses, - the dashed lines are a "*best fit*" through the forward recurrence.

and chose $n = N$ large enough that we can approximate $I_N \approx 0$ then we can generate the smaller n values using the backward recurrence relation. Figure 1.1 also shows the estimate for I_n got using the backward formula (1.5). The results are almost indistinguishable from those of the "*exact*" numerical integration (all the calculations shown are in single precision and the maximum N for the backward recurrence was taken to be 35).

1.3   PROGRAMMING LANGUAGES

Once you have mastered the numerical methods in this book you will need to translate the mathematical approach into a "*programming language*." A programming language is a formal language, which comprises a set of instructions that produce various kinds of output. There is an embarrassment of languages available with new ones begin produced and "*old*" ones modified all the time. I have found some more useful than others.

- Fortran has been in constant use in computationally intensive areas for over seven decades during that time it has evolved, adding extensions and refinements, while striving to retain compatibility with prior versions. "*Modern Fortran*" (Fortran 90/95/03/08) is still the dominant language for the large-scale simulation of physical systems, for things like the astrophysical modeling of stars and galaxies, for the accurate calculation of electronic structure, hydrodynamics, molecular dynamics, and climate change. In the field of high performance computing (HPC), Modern Fortran also has a feature called "*coarrays*" which puts parallelization features directly into the language. Coarrays started as an extension of Fortran 95 and were incorporated into Fortran 2008 as standard. There is a hugh "*legacy*" of libraries both general, e.g., [1–3] and free academic libraries devoted to specific areas in the physical sciences, e.g., [4–8].

- C^{++} is more difficult to learn. It does have a good basis of libraries. On most bench mark tests C^{++} and Fortran are fairly equivalent. However, the two benchmarks where Fortran wins (n-body simulation and calculation of spectra) are the most relevant to Physics.

In the physical sciences C^{++} and "*Modern Fortran*" are still the most widely used. The popular "*Open MPI*" libraries for parallelizing code were developed for these two languages. There is also

- C [9] which was designed to be compiled using a relatively straightforward compiler to provide low-level access to memory and language constructs that map efficiently to machine instructions, all with minimal runtime support. Despite its low-level capabilities, the language was designed to encourage cross-platform programming. A standards-compliant C program written with portability in mind can be compiled for a wide variety of computer platforms and operating systems with few changes to its source code. The language is available on various platforms, from embedded microcontrollers to supercomputers. It is much easier to learn to code in than C^{++}, which is not a direct extension of C. The reason for the name was that when object-oriented languages became popular, C^{++} was originally implemented as a source-to-source compiler; the source code was translated into C, and then compiled with a C compiler.

- Python [10] is easy to learn with built in libraries but it is usually about 100 times slower than Fortran or C^{++} on bench mark tests. A good learning language but not currently really that useful for advanced (real) physical problems.

The challenges you will face in first writing effective code have more to do with knowing how to recast the equations of mathematical physics in a way that minimizes numerical error than with the choice of which high level language in which you are comfortable coding.

CHAPTER 2

Some Elementary Results

In this chapter I give a brief introduction to some of the more basic ideas from numerical analysis.

2.1 TAYLOR'S SERIES

Very often in physical problems you need to find a relatively simple approximation to a complex function or you need to estimate the size of a function. One of the most commonly used techniques is to approximate a function by a polynomial around a given point.

Theorem 2.1 *Let f be a real function which is continuous and has continuous derivatives up to the $n + 1$ order then*

$$f(x) \ = \ f(a) + \frac{f'(a)}{1!}(x - a) + \frac{f^{(2)}(a)}{2!}(x - a)^2 + \cdots + \frac{f^{(n)}(a)}{n!}(x - a)^n + R_n(x),$$

(2.1)

where $n! = n.(n - 1).(n - 2) \ldots 3 \cdot 2 \cdot 1$ and

$$R_n(x) = \int_a^x \frac{f^{(n+1)}(t)}{n!}(x - t)^n \, dt.$$

(2.2)

Proof. [11]. □

Clearly, if R_n goes to zero uniformly as $n \to \infty$ then we can find an infinite series. Examples are

$$\begin{aligned} e^x &= 1 + x + \frac{x^2}{2!} + \cdots, \\ \sin x &= 1 - x + \frac{x^3}{3!} + \cdots. \end{aligned}$$

An alternative form for the remainder term can be derived by by making use of the mean value theorem for integrals, i.e.,

$$R_{n+1} = \int_a^x \frac{f^{(n+1)}(t)}{n!}(x - t)^n \, dt = f^{(n+1)}(\alpha)\frac{(x - a)(x - \alpha)^n}{n!},$$

(2.3)

where α is some number, $a \leq \alpha \leq x$. The form (2.3) is the Cauchy form of the remainder term. A further alternative form was derived by Lagrange

$$R_{n+1}(x) = \frac{f^{(n+1)}(\beta)}{(n+1)!}(x-a)^{n+1}, \tag{2.4}$$

with $a \leq \beta \leq x$. The theorem may be extended to N dimensions.

Theorem 2.2   *Suppose f is a map from $\mathbb{R}^N$ to $\mathbb{R}$ and it is at least $k+1$ times continuously differentiable then:*

$$f(a+h) = \sum_{j=0}^{k} \frac{(h.\nabla)^j}{j!} f(a) + R(a, k, h),$$

$$R(a, k, h) = \frac{1}{(k+1)!}(h.\nabla)^{k+1} f(a+\theta h), \tag{2.5}$$

for some $\theta \in (0, 1)$.

Proof. [12].   □

Example 2.3   The Taylor theorem in two dimensions. Expanding about $a = (a, b)$

$$\begin{aligned}
f(x, y) = \; & f(a, b) + (x-a)\frac{\partial f}{\partial x} + (y-b)\frac{\partial f}{\partial y} \\
& + \frac{1}{(2!)}\left[(x-a)^2\frac{\partial^2 f}{\partial x^2} + 2(x-a)(y-b)\frac{\partial^2 f}{\partial x \partial y} + (y-b)^2\frac{\partial^2 f}{\partial y^2}\right] + \cdots
\end{aligned} \tag{2.6}$$

with all derivatives evaluated at (a, b).

2.1.1   EXTREMA

Suppose $F(x)$ is a continuous function with a continuous first derivate. Suppose further that F has a local maximum at some point x_0, hence for some infinitesimal increment $|h|$

$$F(x_0 \pm |h|) < F(x_0), \tag{2.7}$$

hence

$$\frac{F(x_0 + |h|) - F(x_0)}{|h|} < 0,$$

$$\frac{F(x_0 - |h|) - F(x_0)}{-|h|} > 0.$$

We can make $|h|$ arbitrarily small, hence when we take limit from the left and right and since we have assumed the derivative is continuous we must have:

$$\frac{dF(x_0)}{dx} \equiv \left.\frac{dF(x)}{dx}\right|_{x_0} = 0. \tag{2.8}$$

Following a similar argument it is immediately obvious that if x_0 corresponds to a minimum (2.8) also holds. Now suppose F is a continuous function with continuous first and second derivatives and that their is a point x_0 in its domain where (2.8) holds. Then, using the Taylor's expansion, (2.1) we have

$$\begin{aligned} F(x) &= F(x_0) + (x - x_0)\frac{dF(x_0)}{dx} + \frac{1}{2}\frac{d^2F(x_0)}{dx^2}(x - x_0)^2 + \mathcal{O}((x - x_0)^3) \\ &= F(x_0) + \frac{1}{2}\frac{d^2F(x_0)}{dx^2}(x - x_0)^2 + \mathcal{O}((x - x_0)^3). \end{aligned} \tag{2.9}$$

The symbol $\mathcal{O}(x^n)$ means terms of order x^n or higher. Now since $(x - x_0)^2 > 0$ and since we can chose x arbitrarily close to x_0 we see at once that

$$F \text{ has a maximum at } x_0 \text{ if } \frac{d^2F(x_0)}{dx^2} < 0.$$

$$F \text{ has a minimum at } x_0 \text{ if } \frac{d^2F(x_0)}{dx^2} > 0.$$

Suppose now $f(x, y)$ is a differentiable function of two variables, $f(x, y) = z$ defines a two dimensional surface in three space. We can think of this surface as being constructed from a series of curves of the form $\Phi(x) = f(x, y_0) = z$ and $\Psi(y) = f(x_0, y) = z$, you might think of lines of latitude and longitude on the earth. Clearly, a necessary and sufficient condition for a maximum, is that both Φ and Ψ have maxima. Thus, the function $f(x, y)$ will have an extrema, maximum or minimum at x_0, y_0 is

$$\frac{\partial f(x_0, y_0)}{\partial x} = \frac{\partial f(x_0, y_0)}{\partial y} = 0, \tag{2.10}$$

but

$$df = \frac{\partial f}{\partial x}dx + \frac{\partial f}{\partial y}dy. \tag{2.11}$$

Hence, the condition for an extrema is:

$$df(x_0, y_0) = 0. \tag{2.12}$$

Suppose we may want to find the extrema of $f(x, y)$ where x, y are related by some extra condition

$$g(x_0, y_0) = 0, \tag{2.13}$$

which we will call a constraint. Since g is constant we have $dg = 0$, hence

$$
\begin{aligned}
df &= \frac{\partial f}{\partial x} dx + \frac{\partial f}{\partial y} dy &= 0, \\
dg &= \frac{\partial g}{\partial x} dx + \frac{\partial g}{\partial y} dy &= 0, \\
&\Rightarrow (df - \lambda dg) &= 0, \tag{2.14}
\end{aligned}
$$

for any λ which is independent of x and y. Now define a new function of the three variables x, y, λ

$$l(x, y, \lambda) = f(x, y) - \lambda g(x, y).$$

The exact value of λ will be determined later. The extrema of this function satisfy

$$
\begin{aligned}
\frac{\partial l}{\partial x} &= \frac{\partial f}{\partial x} - \lambda \frac{\partial g}{\partial x} = 0. \\
\frac{\partial l}{\partial y} &= \frac{\partial f}{\partial y} - \lambda \frac{\partial g}{\partial y} = 0. \\
\frac{\partial l}{\partial \lambda} &= -g(x, y) = 0. \tag{2.15}
\end{aligned}
$$

The first two equations must be satisfied by the extrema and the third is just the constraint condition. We have three equations for three unknowns thus the extrema of $f(x, y) = c$ subject to the constraint $g = 0$ can be found by solving for the extrema of the function $l(x, y, \lambda)$.

Example 2.4   Suppose you want to find the maximum and minimum values of the function

$$f(x, y) = xy,$$

where

$$x^2 + y^2 = 4.$$

Then you could proceed by direct substitution and look for the root of $f'(y_0) = 0$, i.e.,

$$
\begin{aligned}
x &= \pm\sqrt{4 - y^2}. \\
f(y) &= x(y)y, \\
f'(y_0) &= \pm \left[\frac{-y^2}{\sqrt{4 - y_0^2}} + \sqrt{4 - y_0^2} \right] = 0,
\end{aligned}
$$

$$\Rightarrow \pm(4 - 2y_0^2) = 0,$$
$$\Rightarrow y_0 = \pm\sqrt{(2)},$$
$$x_0^2 + y_0^2 = 4,$$
$$\Rightarrow x_0 = \pm\sqrt{(2)}.$$

Let us solve the problem using the Lagrange multiplier method

$$l(x, y, \lambda) = xy - \lambda(x^2 + y^2 - 4),$$
$$\frac{\partial l}{\partial x} = y - 2\lambda x = 0,$$
$$\frac{\partial l}{\partial y} = x - 2\lambda y = 0,$$
$$\frac{\partial l}{\partial \lambda} = -(x^2 + y^2 - 4) = 0,$$
$$\Rightarrow \lambda = \frac{y}{2x},$$
$$\Rightarrow x^2 = y^2,$$
$$\Rightarrow x = \pm\sqrt{2},$$
$$y = \pm\sqrt{2},$$

as before.

One of the advantages of the Lagrange multiplier approach is that it easily generalizes to higher dimensions.

Example 2.5   Suppose you wish to find the maximum and minimum values of

$$f(x, y, z) = xyz,$$

on the sphere

$$x^2 + y^2 + z^2 = 3.$$

Let

$$g(x, y, z) = x^2 + y^2 + z^2 - 3.$$

So our constraint is $g(x, y, z) = 0$.
Define

$$l(x, y, z, \lambda) = f(x, y, z) - \lambda g(x, y, z).$$

Table 2.1: Possible values

x	y	z	$f(x,y)$
1	1	1	1
-1	1	1	-1
1	-1	1	-1
1	1	-1	-1
-1	-1	1	1
1	-1	-1	1
-1	-1	1	1
-1	-1	-1	-1

Hence for an extrema,

$$
\begin{aligned}
\frac{\partial l}{\partial x} &= yz - 2x\lambda = 0, \\
\frac{\partial l}{\partial y} &= xz - 2y\lambda = 0, \\
\frac{\partial l}{\partial z} &= xy - 2z\lambda = 0, \\
\frac{\partial l}{\partial \lambda} &= x^2 + y^2 + z^2 - 3 = 0.
\end{aligned}
$$

$$
\begin{aligned}
xyz &= \lambda(2x^2) = \lambda(2y^2) = \lambda(2z^2), \\
\Rightarrow 3xyz &= 2\lambda(x^2 + y^2 + z^2), \\
&= 6\lambda, \\
\Rightarrow xyz &= 2\lambda = 2\lambda x^2, \\
\Rightarrow x &= \pm 1.
\end{aligned}
$$

In the same way $y = \pm 1, z = \pm 1$. We have 8 possible results for extrema.
    So minimum value is -1, maximum is $+1$.

2.1.2   POWER SERIES

Definition 2.6   A power series is a function of a variable x defined as an infinite sum

$$
\sum_{n=0}^{\infty} a_n x^n, \tag{2.16}
$$

where a_n are real numbers.

Just because we write down a series of the form (2.16) it does not mean that such a thing is well defined. It is, in essence, a limit of the sequence of partial sums and this limit may or may not exist. The interval of convergence is the range of values $a < x < b$ for which (2.16) converges. Note this is an open interval that is to say we need to consider the end points separately. If a function f has a Taylor expansion with remainder term R_n which uniformly goes to zero on some interval $I = \{x | a < x < b\}$ then f can be represented by a power series on this interval. Power series are extremely useful. Here we will only state some results and refer the reader to [12] for proof; see also [13].

- A power series may be differentiated or integrated term by term: the resulting series converges to the derivative or the integral of the function represented by the original series within the same interval of convergence.

- Two power series may be added subtracted or multiplied by a constant and this will converge at least within the same interval of convergence, i.e., suppose

$$s_1(x) = \sum_{n=0}^{\infty} a_n x^n,$$
$$s_2(x) = \sum_{m=0}^{\infty} b_m x^m$$

are both convergent within the interval I and α, β are numbers, then

$$s_3(x) = \sum_{n=0}^{\infty} (\alpha a_n + \beta b_n) x^n$$

is convergent within the interval I and

$$s_3(x) = \alpha s_1(x) + \beta s_2(x).$$

- The power series of a function is unique, i.e., if

$$f(x) = \sum_{n=0}^{\infty} a_n x^n,$$
$$f(x) = \sum_{m=0}^{\infty} b_m x^m,$$

then $a_n = b_n$ for all n.

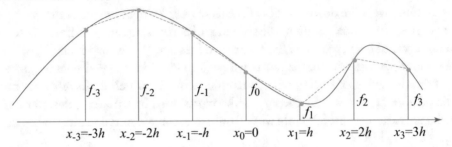

Figure 2.1: Values of f on an equally spaced lattice. Dashed lines show the linear interpolation.

2.2   NUMERICAL DIFFERENTIATION AND INTEGRATION

2.2.1   DERIVATIVES

Suppose f is a known multiply differentiable function defined on some real interval $[a, b]$. We want to find $f'(0)$, the derivative at $x = 0$. Let us suppose we know f on an equally spaced lattice of x values

$$
\begin{aligned}
f_n &= f(x_n), \\
x_n &= nh, (n = 0, \pm 1, \pm 2, \ldots).
\end{aligned}
$$

Using Taylor's theorem

$$
f(x) = f_0 + xf' + \frac{x^2}{2!}f'' + \frac{x^3}{3!}f''' + \cdots , \tag{2.17}
$$

where all derivatives are evaluated at $x = 0$. It follows that

$$
\begin{aligned}
f_{\pm 1} &= f_0 \pm hf' + \frac{h^2}{2}f'' \pm \frac{h^3 f'''}{3!} + \mathcal{O}(h^4), \\
f_{\pm 2} &\equiv f_0 \pm 2hf' + 2h^2 f' \pm \frac{4h^3}{3}f''' + \mathcal{O}(h^4).
\end{aligned} \tag{2.18}
$$

Subtracting f_{-1} from f_1 we find

$$
f' = \frac{f_1 - f_{-1}}{2h} - \frac{h^2}{6}f''' + \mathcal{O}(h^4). \tag{2.19}
$$

The term involving f''' is the dominant error error associated with the finite difference approximation that retains only the first term

$$
f' \approx \frac{f_1 - f_{-1}}{2h}. \tag{2.20}
$$

This "*3-point formula*" will be exact if f is a second degree polynomial. Note also that the symmetric difference about $x = 0$ is used as it is more accurate by one order in h than the forward or backward difference formulae

$$f' \approx \frac{f_1 - f_0}{h} + \mathcal{O}(h),$$

$$f' \approx \frac{f_0 - f_{-1}}{h} + \mathcal{O}(h). \tag{2.21}$$

These "*2-point*" formulae will be exact if f is a linear function on $[0, \pm h]$.

It is possible to improve the 3-point formula, (2.20), by relating f' to lattice points further removed. For example the "*5-point formula*"

$$f' \approx \frac{1}{12h}[f_{-2} - 8f_{-1} + 8f_1 - f_2] + \mathcal{O}(h^5) \tag{2.22}$$

cancels all derivatives in the Taylor series through to fourth order. This formula will be exact f is a fourth-degree polynomial over the 5-point interval $[-2h, 2h]$.

Formulae for higher derivatives can be constructed by taking approximate combinations of (2.18). For example,

$$f_1 - 2f_0 + f_{-1} = h^2 f'' + \mathcal{O}(h^4), \tag{2.23}$$

which leads to the approximation

$$f'' \approx \frac{f_1 - 2f_0 + f_{-1}}{h^2}. \tag{2.24}$$

Numerical differentiation can be quite tricky to program, since by its very nature it involves the subtracting of two very similar numbers.

2.2.2   QUADRATURE

In quadrature we are interested in calculating the definite integral of a function f between two limits $a < b$. We can divide the range.

$$h = \frac{b - a}{N},$$

where N is an integer.

It is then sufficient to derive a formula for the integral from $-h$ to h since this formula can then be applied successively

$$\int_a^b f(x)dx = \int_a^{a+2h} f(x)dx + \int_{a+2h}^{a+4h} f(x)dx + \cdots + \int_{b-2h}^b f(x)dx. \tag{2.25}$$

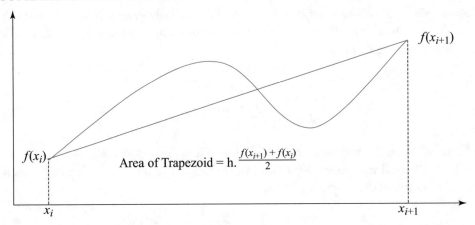

Figure 2.2: Using the trapezoidal rule to integrate $\int_{x_n}^{x_{n+1}} f(x)dx$ corresponds to approximating the integral by the area defined by the right trapezoid whose area is the sum of the rectangle $h \times f(x_n)$ with the right triangle whose area is $\frac{1}{2}(f(x_{n+1}) - f(x_n))h$, in agreement with (2.26).

The idea is to approximate f on each interval by a function that is integrable, this approach leads to a group of formulae that are said to be of "*Newton–Cotes*" type. Let us first consider $\int_{-h}^{h} f(x)dx$.

If $f(x)$ is a linear function then

$$\int_{-h}^{h} f(x)dx = \frac{h}{2}(f_{-1} + 2f_0 + f_1) \tag{2.26}$$

is exact on $[-h, h]$. Thus, from Taylor's theorem it follows that

$$\int_{-h}^{h} f(x)dx = \frac{h}{2}(f_{-1} + 2f_0 + f_1) + \mathcal{O}(h^3). \tag{2.27}$$

The approximation (2.26) is known as the "*trapezoidal rule*."

Lemma 2.7 *For poylnomials of order 3 or less*

$$\int_{-h}^{h} f(x)dx = \frac{h}{3}[f(+h) + 4f(0) + f(-h)]. \tag{2.28}$$

Proof.

$$\int_{-h}^{h} 1dx = x|_{-h}^{h} = 2h = \frac{h}{3}[1 + 4 + 1] = 2h,$$

$$\int_{-h}^{h} x\,dx = \frac{1}{2}x^2|_{-h}^{h} = 0 = \frac{h}{3}[1 + 0 - 1] = 0,$$

$$\int_{-h}^{h} x^2\,dx = \frac{x^3}{3}|_{-h}^{h} = \frac{2h^3}{3} = \frac{h}{3}\left[h^2 + 0 + h^2\right] = \frac{2h^3}{3},$$

$$\int_{-h}^{h} x^3\,dx = \frac{x^4}{4}|_{-h}^{h} = 0 = \frac{h}{3}\left[h^3 + 0 - h^3\right] = 0.$$

$\square$

Suppose we want to find the integral

$$\int_{a}^{b} f(x)\,dx.$$

We chose to work with an *even* number of lattice spacings

$$N = \frac{(b-a)}{h}.$$

We can write the integral as

$$\int_{a}^{b} f(x)\,dx = \int_{a}^{a+2h} f(x)\,dx + \int_{a+2h}^{a+4h} f(x)\,dx + \cdots + \int_{b-2h}^{b} f(x)\,dx. \qquad (2.29)$$

Consider the first and second integrals on the right:

$$\int_{a}^{a+2h} f(x)\,dx = \int_{-h}^{h} f(u)\,du, \; u = x - a - h,$$

$$\int_{a+2h}^{a+4h} f(x)\,dx = \int_{-h}^{h} f(z)\,dz, \; z = x - a - 3h. \qquad (2.30)$$

Now applying Lemma 2.7 to both integrals we have

$$\int_{a}^{a+2h} f(x)\,dx \approx \frac{h}{3}[f(u = h) + 4f(u = 0) + f(u = h)]$$

$$= \frac{h}{3}[f(x = a + 2h) + 4f(x = a + h) + f(x = a)],$$

$$\int_{a+2h}^{a+4h} f(x)\,dx \approx \frac{h}{3}[f(z = h) + 4f(z = 0) + f(z = h)]$$

$$= \frac{h}{3}[f(a + 4h) + 4f(a + h) + f(a + 2h)].$$

Continuing in this way we eventually get

$$\int_{a}^{b} f(x)\,dx = \frac{h}{3}[f(a) + 4f(a + h) + 2f(a + 2h) + 4f(a + 3h)$$

$$+ \cdots + 4f(b-h) + f(b)]. \tag{2.31}$$

This is "*Simpson's rule*" which is accurate to two orders better than the trapizoidal rule.

High order quadrature formulas can be derived by retaining more terms in the Taylor's expansion used to interpolate f and using better finite difference approximations for the derivatives. The generalization of Simpsons rule using cubic and quartic polynomials are

Simpson's $\frac{3}{8}$ rule

$$\int_{x_0}^{x_3} f(x)dx = \frac{3h}{8} [f_0 + 3f_1 + 3f_2 + f_3] + \mathcal{O}(h^5). \tag{2.32}$$

Boole's Rule

$$\int_{x_0}^{x_4} f(x)dx = \frac{2h}{45} [7f_0 + 32f_1 + 12f_2 + 32f_3 + 7f_4] + \mathcal{O}(h^7). \tag{2.33}$$

Note[1] for this method to be applicable N must be a multiple of 4.

Although one might think that formulae based on interpolation using polynomials of a very high degree would be even more accurate this is not necessarily the case since.

 (i) Such polynomials tend to oscillate violently and thus lead to inaccurate interpolation.

 (ii) The coefficients of the values of f can have both positive and negative signs in higher order, making cancellation errors a potential problem.

Later on we will derive quadrature formulae which are accurate to a higher order but to do this we will have to agree to give up on having equally spaced abscissae.

2.2.3 SINGULAR INTEGRANDS, INFINITE INTEGRALS

We need to be careful if our integrand is singular, even if the actual integral is well defined. Sometimes the best approach is just to make a change of variable.

Example 2.8

$$I_1 = \int_0^1 x^{-1/3} g(x)dx. \tag{2.34}$$

In this case put

$$t^3 = x,$$

---

[1] Due to a misprint in [14]. This approximation was incorrectly written as "*Bode's rule*," this error is frequently reproduced in the literature.

$$\Rightarrow I_1 = 3 \int_0^1 t^{-1} t^2 g(t^3) dt$$

$$= 3 \int_0^1 t g(t^2) dt. \tag{2.35}$$

In some cases it is better to isolate the singularity; for example,

$$\int_0^1 \frac{\sin(x)}{x} dx = \int_0^h \frac{\sin(x)}{x} dx + \int_h^1 \frac{\sin(x)}{x} dx$$

$$= \int_0^h \left(-\frac{x^3}{6} + \mathcal{O}(x^5) \right) dx + \int_h^1 \frac{\sin(x)}{x} dx. \tag{2.36}$$

The Simpson's rule calculation of

$$\int_1^\infty x^{-2} g(x) dx$$

is not defined. However, changing variables

$$x = t^{-1}$$

gives

$$\int_0^1 d g(t^{-1}) dt,$$

which can be evaluated with any of the formulae we discussed.

2.3 FINDING ROOTS

We will frequently need to find the roots of a function f that is we will want to find the values of x_0, s.t.

$$f(x_0) = 0.$$

Given a continuous function f defined on $[a_0, b_0]$, suppose

$$f(a_0) f(b_0) < 0;$$

then f must have at least one root in (a_0, b_0). The trick here is to repeatedly bisect, all the time decreasing the size of the interval. Define

$$c = \frac{1}{2} (a_0 + b_0).$$

If $f(c)$ has the same sign as $f(b)$ then must be a root in $[a_0, c]$ otherwise there is a root in $[c, b_0]$. Either way we have halved the size of the interval containing the root. The rule we adopt, starting with $N = 0$, is:

$$\text{If } f(b_N)f(c) \; > \; 0 \text{ then take } a_{N+1} = a_N, b_{N+1} = c,$$
$$\text{If } f(b_N)f(c) \; < \; 0 \text{ then take } a_{N+1} = c, b_{N+1} = b_N,$$
$$c \; = \; \frac{1}{2}\left(|b_{N+1} - a_{N+1}|\right).$$

After N iterations the position of the root is no more than

$$\frac{1}{2}|b_N - a_N|$$

from the mid-point of the interval $[a_N, b_N]$. But

$$\frac{1}{2}|b_N - a_N| = \frac{1}{2^{N+1}}|b_0 - a_0|.$$

This means that after N iterations the root must lie in an interval that is no bigger than

$$\epsilon \; = \; \frac{1}{2^{N+1}}|b_0 - a_0|.$$

We can chose ϵ to the desired tolerance then we know that we need at most N iterations where

$$N = \frac{\ln|b_0 - a_0| - \ln(\epsilon)}{\ln 2} - 1$$

to converge to within this tolerance.

As an example, I looked for the solution to

$$f(x) = x^2 - 5 = 0, \tag{2.37}$$

with a tolerance of 10^{-6} using $x = 1$ as my initial guess and an initial step size of 0.5, and my code converged to the answer, correct to 6 places of decimals, after 34 iterations. You need to be careful using this method, since if the initial step size is too large it is possible to "*step over*" the desired root especially when f has several roots.

If the actual root is at x_0 and we guess x_1. If it is a good guess $|x_0 - x_1|$ will be small. Using Taylor's

$$f(x_0) \; = \; 0$$
$$= \; f(x_1) + f'(x_1)(x_0 - x_1) + \mathcal{O}((x_0 - x_1)^2).$$
$$\Rightarrow x_0 \; \approx \; x_1 - \frac{f(x_1)}{f'(x_1)}.$$

Thus, a better guess for the root will be

$$x_2 = x_1 - \frac{f(x_1)}{f'(x_1)}.$$

Repeating, we get

$$x_{i+1} = x_i - \frac{f(x_i)}{f'(x_i)}. \tag{2.38}$$

The application of (2.38) defines the Newton–Raphson algorithm. I used it to look for the root of $x^2 - 5 = 0$ with a tolerance of 10^{-6}. I was able to achieve convergence after only 10 iterations.

The "*secant method*" is useful if finding the derivative is a problem.

We approximate

$$f'(x_i) \approx \frac{f(x_i) - f(x_{i-1})}{x_i - x_{i-1}},$$

and rewrite (2.38)

$$x_{i+1} = x_i - f(x_i) \frac{x_i - x_{i-1}}{f(x_i) - f(x_{i-1})}. \tag{2.39}$$

Provide that the initial guesses are reasonably close to the true root, convergence to the exact answer is almost as rapid as the Newton–Raphson algorithm. The Newton–Raphson and secant methods can fail to converge or worse converge to the wrong answer if there are multiple roots close together or if there is a point, $\tilde{x}$ near x_0 where $f'(\tilde{x}) = 0$.

CHAPTER 3

The Numerical Solution of Ordinary Differential Equations

It is sometimes, misleadingly, said that *"physicists can't solve the few body problem,"* when what is really true is that there is *"no general analytical solution to the few body problem given by simple algebraic expressions."* Sometimes it is nice to see a solution to an important problem written down in terms of the elementary functions known to the ancients. However, it is often enough to have a well-posed differential equation to find all that we could possible want to know about a physical system. I will illustrate this by considering the most well known of all analytic functions [11, 12].

3.1 TRIGONOMETRIC FUNCTIONS

Lemma 3.1 *Let $c(x), s(x)$ be continuous differentiable functions such that*

$$
\begin{aligned}
s'(x) &= c(x), \\
c'(x) &= -s(x), \\
s(0) &= 0, \\
c(0) &= 1.
\end{aligned}
\tag{3.1}
$$

Then

$$
c^2(x) + s^2(x) = 1.
\tag{3.2}
$$

Proof. Let

$$
\begin{aligned}
F(x) &= c^2(x) + s^2(x), \\
F'(x) &= 2c(x)c'(x) + 2s(x)s'(x) = 0,
\end{aligned}
$$

thus $F(x)$ must be a constant, substituting the values at $x = 0$ we have the result. $\qquad\square$

Lemma 3.2   *If we have two sets of functions $c(x), s(x)$, and $f(x), g(x)$ s.t.,*

$$
\begin{aligned}
c'(x) &= -s(x), & g'(x) &= -f(x), \\
s'(x) &= c(x), & f'(x) &= g(x), \\
c(0) &= 1, & g(0) &= 1, \\
s(0) &= 0, & f(0) &= 0.
\end{aligned}
$$

Then $f(x) = s(x); c(x) = g(x)$ for all x.

Proof. We know that both the pairs $(c, s), (f, g)$ must satisfy the relation (3.2)

$$
\begin{aligned}
c^2(x) + s^2(x) &= 1, \\
f^2(x) + g^2(x) &= 1.
\end{aligned}
$$

The functions
$F_1(x) = f(x)c(x) - s(x)g(x)$ and $F_2(x) = f(x)s(x) + c(x)g(x)$ are s.t.

$$
\frac{dF_1(x)}{dx} = \frac{dF_2(x)}{dx} = 0.
$$

Hence,

$$
\begin{aligned}
a &= f(x)c(x) - s(x)g(x), \\
b &= f(x)s(x) + c(x)g(x),
\end{aligned}
$$

where a, b are constants putting in the values at $x = 0$ yields:

$$
\begin{aligned}
0 &= f(x)c(x) - s(x)g(x), \\
1 &= f(x)s(x) + c(x)g(x);
\end{aligned}
$$

hence,

$$
\begin{aligned}
0 &= f(x)c^2(x) - c(x)s(x)g(x), \\
s(x) &= f(x)s^2(x) + s(x)c(x)g(x),
\end{aligned}
$$

adding the last two lines yields

$$
s(x) = f(x).
$$

Hence,

$$
s'(x) = f'(x);
$$

therefore,

$$
c(x) = g(x).
$$

$\square$

Clearly, the functions $c(x), s(x)$ have all the properties of the $\sin(x)$ and $\cos(x)$ of trigonometry. The rest of the properties that we know and love can be derived from the results above. Further, as we will see slightly later, we can use relatively straight forward numerical methods to solve equations of the form (3.1). This may seem an odd way to discuss the sin and cos functions but there is an important lesson here, in that perfectly good functions can be defined simply as the solution of differential equations. The Schrödinger equation for an N electron neutral atom in atomic units

$$\left[\sum_{j=1}^{N} \left[-\frac{1}{2} \nabla_j^2 - \frac{Z}{r_j} \right] + \frac{1}{2} \sum_{j \neq k}^{N} \frac{1}{|r_j - r_k|} - E \right] \Psi(r_1, r_2, \ldots, r_N) = 0,$$

is just another differential equation, albeit a more complicated one, which turns out to have a unique solution for certain values of E.

3.2   ANALYTIC SOLUTIONS

Suppose we are presented with the O.D.E.

$$\frac{dy(t)}{dt} + p(t)y(t) = g(t). \tag{3.3}$$

Then there is a convenient trick. Define, with a a constant,

$$r(t) = \exp\left(\int_a^t p(x)dx \right),$$
$$\Rightarrow \frac{dr}{dt} = \frac{de^u}{du} \frac{du}{dt} \quad \text{where} \quad u(t) = \int_a^t p(x)dx,$$
$$\frac{dr}{dt} = e^u p(t),$$
$$= e^{\int_a^t p(x)dx} p(t). \tag{3.4}$$

Returning to (3.3) and multiplying by $\int_a^t p(x)dx$ we have

$$\frac{dr(t)y(t)}{dt} = g(t)r(t). \tag{3.5}$$

Example 3.3   Suppose we want to solve

$$t\frac{dy(t)}{dt} + 2y = 4t^2 \tag{3.6}$$

subject to

$$y(1) = 2.$$

Then we can devide (3.6) by t to put it in the form (3.4):

$$\frac{dy(t)}{dt} + 2\frac{y}{t} = 4t,$$

$$r(t) = \exp\left(\int_a^t \frac{2}{x} dx\right),$$

$$\Rightarrow r(t) = \exp(2\ln t - 2\ln a), \tag{3.7}$$

with out loss of generality you can take the arbitrary constant a to be unity and then

$$r(t) = \exp(2\ln t) = \exp(\ln t^2) = t^2 \tag{3.8}$$

if we multiply (3.6) by t^2 we have

$$t^2 \frac{dy}{dt} + 2ty = 4t^3,$$

$$\Rightarrow \frac{dt^2 y}{dt} = 4t^3,$$

$$\Rightarrow t^2 y(t) = t^4 + c, \tag{3.9}$$

where c is a constant now substitute the initial condition $y(1) = 2$ and it follows that $c = 1$.

Some times we can use direct integration. Consider

$$x\frac{dy}{dx} = y\ln y$$

$$y(2) = e. \tag{3.10}$$

We can rewrite

$$\frac{dy}{y\ln y} = \frac{dx}{x},$$

$$\int \frac{dy}{y\ln y} = \int \frac{dx}{x}$$

$$= \ln(x) + K. \tag{3.11}$$

Put

$$u = \ln y,$$

$$du = \frac{dy}{y},$$

$$\Rightarrow \int \frac{dy}{y\ln y} = \int \frac{du}{u},$$

$$= \ln u,$$

$$= \ln(\ln(y)),$$

$$= \ln(x) + K. \tag{3.12}$$

Initial conditions

$$\begin{aligned}
y(2) &= e \\
\Rightarrow \ln(\ln(e)) &= \ln(2) + K, \\
\ln(1) &= \ln(2) + K, \\
K &= -\ln(2).
\end{aligned} \tag{3.13}$$

So from (3.12),

$$\begin{aligned}
\ln(\ln(y)) &= \ln(x) - \ln(2), \\
&= \ln(\frac{x}{2}), \\
\Rightarrow y &= e^{\frac{x}{2}}.
\end{aligned} \tag{3.14}$$

3.3   NUMERICAL METHODS

3.3.1   EULER APPROXIMATION

Consider the ordinary first-order differential equation

$$\frac{dx(t)}{dt} = f(t, x), \tag{3.15}$$

with initial condition $x(t_0) = x_0$. Now, from Taylor's theorem

$$\begin{aligned}
x(t + h) &= x(0) + h\frac{dx}{dt}|_{t=t_0} + \mathcal{O}(h^2), \\
&\approx x_0 + hf(t_0, x_0).
\end{aligned} \tag{3.16}$$

Equation (3.16) is known as the Euler solution. We could use it to propagate our solution using a series of increments in h; at each step we introduce a potential error of order h^2.

Example 3.4   Let us apply the Euler method to (3.10)

$$\begin{aligned}
x\frac{dy}{dx} &= y \ln y, \\
y(2) &= e,
\end{aligned}$$

whose solution we know to be $y = e^{\frac{x}{2}}$. Let us take $h = 0.1$

$$\begin{aligned}
y(x + h) &\approx y(x) + hf(x, y), \\
x_0 &= 2, \\
y_0 &= e, \\
f(x, y) &= \frac{y \ln y}{x}.
\end{aligned}$$

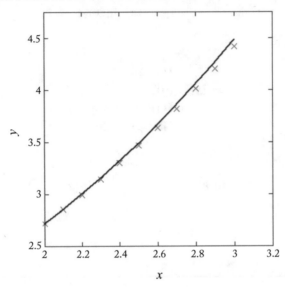

Figure 3.1: Euler solution, crosses compared with analytic result, solid line.

Step 1.

$$
\begin{aligned}
y_1 = y(2.1), \quad &= \quad y_0 + hf(x_0, y_0), \\
&= \quad e + hf(2, e), \\
&= \quad e + 0.1 \times \frac{e \ln e}{2}, \\
&= \quad 2.85419583.
\end{aligned}
$$

Step 2.

$$
\begin{aligned}
y_2 = y(2.2), \quad &= \quad y_1 + hf(x_1, y_1), \\
&= \quad 2.85419583 + h \times f(2.1, 2.85419583), \\
&= \quad 2.99674129.
\end{aligned}
$$

It is straightforward to write a computer program to propagate the solution further. In Figure 3.1, I show a comparison between the results of my numerical code and the analytic solution.

We can generalize to higher order differential equations.

Example 3.5   Consider the classical harmonic oscillator problem

$$
\frac{d^2 x}{dt^2} \quad = \quad -x,
$$

$$\frac{dx(0)}{dt} = 0,$$
$$x(0) = 1. \tag{3.17}$$

We can convert this into a set of two first-order equations

$$\frac{dv}{dt} = -x,$$
$$\frac{dx}{dt} = v,$$
$$v(0) = 0,$$
$$x(0) = 1. \tag{3.18}$$

We can apply Taylor's theorem to both $x(t)$ and $v(t)$

$$v(t+h) = v(0) + h\frac{dv(0)}{dt},$$
$$= v(0) - hx(0),$$
$$x(t+h) = x(0) + h\frac{dx(0)}{dt},$$
$$= x(0) + hv(0); \tag{3.19}$$

incrementing we get

$$y_{N+1} = \begin{pmatrix} x \\ v \end{pmatrix}\Big|_{N+1} = \begin{pmatrix} x \\ v \end{pmatrix}\Big|_N + h\begin{pmatrix} \frac{dx}{dt} \\ \frac{dv}{dt} \end{pmatrix}_N$$
$$= y_N + h\begin{pmatrix} v \\ -x \end{pmatrix}\Big|_N. \tag{3.20}$$

To code this up we can divide the interval $[0, \pi]$ into h equal steps and create two arrays $x(0:100)$, $v(0, 100)$ and then iterate.

In Figure 3.2, I show the Euler numerical solution from (3.20) plotted against $\cos t$ the analytic solution to (3.17).

3.3.2 RUNGE–KUTTA METHOD

The second order Runge–Kutta is most simply derived by applying the trapezoidal rule to integrating

$$\frac{dy}{dt} = f(y, t), \tag{3.21}$$

over the interval $[t_n, t_{n+1}]$.

$$y_{n+1} = y_n + \int_{t_n}^{t_{n+1}} f(y, t)dt,$$

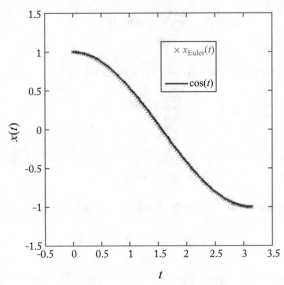

Figure 3.2: Comparison of the analytic solution to (3.17), solid line, and the numerical solution found using the Euler approach, crosses.

$$\approx \quad y_n + \frac{h}{2}\left[f(y_n, t_n) + f(\bar{y}_{n+1}, t_{n+1})\right].$$

We approximate $\bar{y}_{n+1}$ using the Euler method

$$y_{n+1} = y_n + \frac{h}{2}\left[f(y_n, t_n) + f(y_n + hf(y_n, t_n), t_{n+1})\right].$$

It is convenient to define

$$
\begin{aligned}
k_1 &= hf(y_n, t_n), \\
k_2 &= hf(y_n + k_1, t_{n+1}), \\
\Rightarrow y_{n+1} &= y_n + \frac{1}{2}\left[k_1 + k_2\right].
\end{aligned}
\tag{3.22}
$$

Equation (3.22) defines the second order Runge–Kutta approximation. We can get a better approximation by improving our estimate for $\bar{y}_{n+1}$. The Runge–Kutta approximation can be extended to higher orders [15]. The fourth order Runge–Kutta is given by

$$
\begin{aligned}
y_{n+1} &= y_n + \frac{1}{6}\left[k_1 + 2k_2 + 2k_3 + k_4\right] + \mathcal{O}(h^5), \\
\text{where} \\
k_1 &= hf(y_n, t_n), \\
k_2 &= hf(y_n + \frac{1}{2}k_1, t_n + \frac{1}{2}h),
\end{aligned}
$$

$$k_3 = hf(y_n + \frac{1}{2}k_2, t_n + \frac{1}{2}h),$$
$$k_4 = hf(y_n + k_3, t_n + h). \tag{3.23}$$

The method can be extended to find the numerical solution to nth order differential equations. In much the same way as we looked at the vector formalism for the Euler method (3.20), we can generalize the Runge–Kutta. For example the second order differential equation:

$$\frac{d^2 f}{dt^2} = g(t)$$
$$f(0) = f_0$$
$$f'(0) = v_0$$

can be transformed into two coupled differential equations:

$$\frac{du_1(t)}{dt} = u_2(t),$$
$$\frac{du_2(t)}{dt} = g(t),$$
$$u_1(0) = f_0,$$
$$u_2(0) = v_0, \tag{3.24}$$

and solved using the vector scheme:

$$\dot{\boldsymbol{u}} = \boldsymbol{f}(t, \boldsymbol{u}) = \begin{bmatrix} u_2 \\ g(t) \end{bmatrix},$$
$$\boldsymbol{k}_1 = h\boldsymbol{f}(t_n, \boldsymbol{u}_n),$$
$$\boldsymbol{k}_2 = h\boldsymbol{f}(t_n + \frac{1}{2}h, \boldsymbol{u}_n + \frac{1}{2}\boldsymbol{k}_1),$$
$$\boldsymbol{k}_3 = h\boldsymbol{f}(t_n + \frac{1}{2}h, \boldsymbol{u}_n + \frac{1}{2}\boldsymbol{k}_2),$$
$$\boldsymbol{k}_4 = h\boldsymbol{f}(t_n + h, \boldsymbol{u}_n + \boldsymbol{k}_3),$$
$$\boldsymbol{u}_{n+1} = \boldsymbol{u}_n + \frac{1}{6}(\boldsymbol{k}_1 + \boldsymbol{k}_2 + \boldsymbol{k}_3 + \boldsymbol{k}_4). \tag{3.25}$$

Second and higher order ordinary differential equations (more generally, systems of nonlinear equations) rarely yield closed form solutions. A great advantage of the numerical approach is that it can be applied to both linear and nonlinear differential equations. A numerical solution to the classical harmonic oscillator problem

$$\frac{d^2 x}{dt^2} = -x,$$
$$x(0) = 1,$$
$$\left. \frac{dx}{dt} \right|_{t=0} = 0, \tag{3.26}$$

using the fourth order Runge–Kutta scheme as given in (3.25) results are shown in Figure 3.3 with $h = 0.3$.

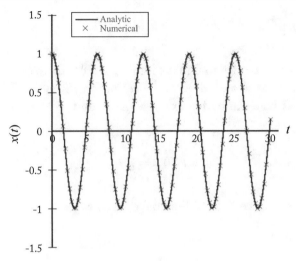

Figure 3.3: Comparison of the analytic solution to (3.26), solid blue line compared to the fourth order Runge–Kutta calculation, $h = 0.3$, red crosses.

3.3.3   NUMEROV METHOD

Suppose we are given the differential equation

$$y''(x) = -g(x)y(x) + s(x). \tag{3.27}$$

To derive the Numerov's method for solving this equation, we begin with the Taylor expansion of the function we want to solve, $y(x)$, around the point x_0

$$y(x) = y(x_0) \quad + \quad (x - x_0)y'(x_0) + \frac{(x - x_0)^2}{2!}y''(x_0) + \frac{(x - x_0)^3}{3!}y'''(x_0)$$
$$+ \quad \frac{(x - x_0)^4}{4!}y''''(x_0) + \frac{(x - x_0)^5}{5!}y'''''(x_0) + \mathcal{O}(h^6). \tag{3.28}$$

Denoting the distance from x to x_0 by $h = x - x_0$, we can write the above equation as

$$y(x_0 + h) \quad = \quad y(x_0) + hy'(x_0) + \frac{h^2}{2!}y''(x_0) + \frac{h^3}{3!}y'''(x_0)$$
$$+ \quad \frac{h^4}{4!}y''''(x_0) + \frac{h^5}{5!}y'''''(x_0) + \mathcal{O}(h^6). \tag{3.29}$$

If we evenly discretize the space, we get a grid of x points, where $h = x_{n+1} - x_n$. By applying the above equations to this discreet space, we get a relation between y_n and y_{n+1}

$$y_{n+1} \quad = \quad y_n + hy'(x_n) + \frac{h^2}{2!}y''(x_n) + \frac{h^3}{3!}y'''(x_n)$$

$$+ \frac{h^4}{4!}y''''(x_n) + \frac{h^5}{5!}y'''''(x_n) + \mathcal{O}(h^6). \tag{3.30}$$

Computationally, this amounts to taking a step *"forward"* by an amount h. If we want to take a step *"backward,"* we replace every h with $-h$ and get the expression for y_{n-1}

$$\begin{aligned}
y_{n-1} &= y_n - hy'(x_n) + \frac{h^2}{2!}y''(x_n) - \frac{h^3}{3!}y'''(x_n) \\
&+ \frac{h^4}{4!}y''''(x_n) - \frac{h^5}{5!}y'''''(x_n) + \mathcal{O}(h^6).
\end{aligned} \tag{3.31}$$

Summing the two equations, we find that

$$y_{n+1} - 2y_n + y_{n-1} = h^2 y_n'' + \frac{h^4}{12}y_n'''' + \mathcal{O}(h^6). \tag{3.32}$$

We can solve this equation for y_{n+1} by substituting the expression given at the beginning, that is $y_n'' = -g_n y_n + s_n$. To get an expression for the y_n'''' factor, we simply have to differentiate

$$y'' = -gy + s \tag{3.33}$$

twice and approximate it again as we did above:

$$\begin{aligned}
y'''' &= \frac{d^2}{dx^2}(-gy), \\
\Rightarrow h^2 y_n'''' &= -g_{n+1}y_{n+1} + s_{n+1} + 2g_n y_n - 2s_n - g_{n-1}y_{n-1} + s_{n-1} + \mathcal{O}(h^4).
\end{aligned} \tag{3.34}$$

If we now substitute this in (3.32), we get

$$\begin{aligned}
y_{n+1} - 2y_n + y_{n-1} &= \\
h^2(-g_n y_n + s_n) \quad &+ \quad \frac{h^2}{12}(-g_{n+1}y_{n+1} + s_{n+1} + 2g_n y_n - 2s_n - g_{n-1}y_{n-1} + s_{n-1}) \\
&+ \quad \mathcal{O}(h^6).
\end{aligned} \tag{3.35}$$

Rearranging

$$y_{n+1}\left(1 + \frac{h^2}{12}g_{n+1}\right) - 2y_n\left(1 - \frac{5h^2}{12}g_n\right) + y_{n-1}\left(1 + \frac{h^2}{12}g_{n-1}\right)$$
$$= \frac{h^2}{12}(s_{n+1} + 10s_n + s_{n-1}) + \mathcal{O}(h^6).$$

$$\Rightarrow y_{n+1} \approx \frac{2y_n\left(1 - \frac{5h^2}{12}g_n\right) - y_{n-1}\left(1 + \frac{h^2}{12}g_{n-1}\right) + \frac{h^2}{12}(s_{n+1} + 10s_n + s_{n-1})}{\left(1 + \frac{h^2}{12}g_{n+1}\right)}.$$

(3.36)

One might expect that the errors at each step would be roughly comparable so the total error in the Numerov method would be $\mathcal{O}(h^6 h^{-1}) = \mathcal{O}(h^5)$. Unfortunately this is generally not true, the error tends to grow with each step and a better estimate is $\mathcal{O}(h^4)$ the same as the 4th order Runge–Kutta. It main disadvantages are that we need both y_0 and y_1 to start it off and that round off errors can pop up when applying (3.36), you should always use double precision in your Numerov code.

CHAPTER 4

Case Study: Damped and Driven Oscillations

4.1 LINEAR AND NONLINEAR ORDINARY DIFFERENTIAL EQUATIONS

The great advantage of using the numerical approach to solving differential equations is that it can be applied equally well to nonlinear as well as linear equations. A differential equation is linear if it involves the dependent variables and their derivatives only linearly. For example, the familiar equation

$$\frac{d^2x}{dt^2} = -\omega_0^2 x(t)$$

is a linear one but

$$\frac{d^2x}{dt^2} = -\omega_0^2 \sin(x(t))$$

is nonlinear.

Linear equations are much simpler to deal with. Consider the general homogenous second order ordinary differential equation

$$a_2(t)\frac{d^2x}{dt^2} + a_1(t)\frac{dx}{dt} + a_0(t)x = 0. \tag{4.1}$$

If x_1 and x_2 are solutions of (4.1) then

$$\alpha x_1(t) + \beta x_2(t)$$

is also a solution. A consequence of this is that if we find two linearly independent solutions, $x_1(t), x_2(t)$ to (4.1) then the general solution is given by $x_g(t) = \alpha x_1(t) + \beta x_2(t)$ where α, β are fixed by the boundary conditions [11, 16]. This useful property does not apply to nonlinear differential equations and consequently very few nonlinear equations admit a simple analytic solution. For a linear equation a small change in initial conditions generally only produces a small change in the final solution but some nonlinear equations can exhibit can extreme sensitivity to initial conditions. The good news is that numerical methods such as the Runge–Kutta can be applied equally effectively to both linear and nonlinear problems. In this chapter, I will illustrate some of these properties by considering the physical pendulum. For a more compete discussion see [17].

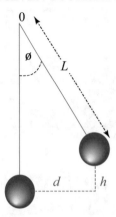

Figure 4.1: The simple pendulum consists of a heavy weight attached to a fixed point by a massless string. The string is of length L. It is displaced from equilibrium through some small angle, ϕ, and allowed to oscillate.

4.2   THE PHYSICAL PENDULUM

Let us consider the pendulum. A heavy mass is hung from a fixed point by a light inextensible string; in other words, we assume the weight of the string is negligible compared to the weight of the mass and the string is strong enough not to stretch when the mass hangs down. Initially, the mass is in equilibrium with the gravitational force being exactly cancelled by the tension in the string. If we displace the mass through an angle ϕ_0 and release it will begin to fall under gravity. At some time t it will make an angle $\phi(t)$ with the vertical. If we assign zero potential energy to the equilibrium point, i.e., hanging straight down with $\phi = 0$ then at time t we have

$$V = mgh, \tag{4.2}$$

where h is the vertical displacement above equilibrium; see Figure 4.1.

Clearly,

$$\begin{aligned} \sin\phi &= \frac{d}{L}, \\ \cos\phi &= \frac{L-h}{L}, \\ \Rightarrow h &= L(1-\cos\phi). \end{aligned} \tag{4.3}$$

The potential energy is then

$$V(\phi) = MgL(1-\cos\phi). \tag{4.4}$$

Now if we think of our polar coordinates being fixed at the origin then [11]

$$v = \dot{r}e_r + r\dot{\phi}e_\phi.$$

In our case $r = L, \dot{r} = 0$; hence, the kinetic energy is

$$K = \frac{1}{2}ML^2\dot{\phi}^2, \tag{4.5}$$

and the total constant energy is

$$
\begin{aligned}
E &= K + V, \\
E &= \frac{1}{2}ML^2\dot{\phi}^2 + \frac{1}{2}MgL\phi^2. \\
\frac{dE}{dt} &= 0, \\
\Rightarrow 0 &= ML^2\ddot{\phi}\dot{\phi} + MgL\dot{\phi}\sin\phi, \\
\Rightarrow \ddot{\phi} &= -\frac{g}{L}\sin(\phi) \\
&= -\omega_0^2\sin(\phi), \tag{4.6}
\end{aligned}
$$

where

$$\omega_0 = \sqrt{\frac{g}{L}}.$$

4.2.1   SMALL OSCILLATIONS

First, let us consider the idealization where ϕ is "small" and $\sin\phi \approx \phi$, then (4.6) reduces to

$$\ddot{\phi} = -\omega_0^2\phi. \tag{4.7}$$

This is a simple harmonic oscillator equation with general solution

$$\phi(t) = A\cos(\omega_0 t + \eta), \tag{4.8}$$

where

$$\omega_0 = \sqrt{\frac{g}{l}}.$$

The constants A and η are determined by our initial conditions. If, for example, the mass is released from rest at $t = 0$ at an angle ϕ_0, then

$$
\begin{aligned}
\dot{\phi}(0) &= 0 \\
&= -A\omega_0\sin(\eta), \\
\Rightarrow \eta &= 0, \\
\Rightarrow A &= \phi_0, \\
\Rightarrow \phi(t) &= \phi_0\cos(\omega_0 t), \\
\dot{\phi} &= -\omega_0\phi_0\sin(\omega_0 t). \tag{4.9}
\end{aligned}
$$

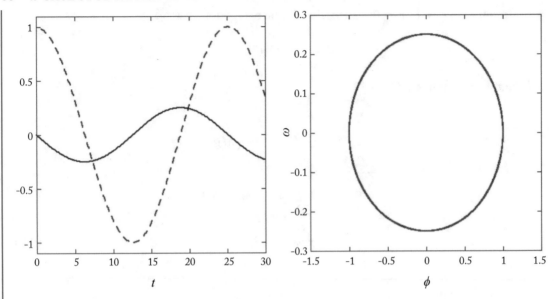

Figure 4.2: The undamped oscillator with $\gamma = 0.0$, $Q = 0.0$ with $\omega_0 = 0.25$ shown in the left panel $\phi(t)$ red dashed, $\omega = \dot{\phi}$, solid blue as a function of time, the right panel shows the phase trajectory ω against ϕ.

The motion repeats itself indefinitely with a period of

$$T = \frac{2\pi}{\omega_0}.$$

Figure 4.2 shows the time dependence of ϕ and $\omega = \dot{\phi}$ as a function of time. Also shown is the phase trajectory where ω is plotted against ϕ. In this simple case, the phase trajectory is an ellipse. At a time $\frac{T}{4}$ after being released the mass will be back at the origin with its maximum speed with all energy kinetic. It will then decelerate and come to a stop at time $t = \frac{T}{2}$ at an angle of $-\phi_0$. At this point all its energy is potential. After a further time of $\frac{T}{4}$ it is back at the origin with only kinetic energy and after a total time T it is back at its original position with $\phi = \phi_0$ and $\dot{\phi} = 0$. This process will repeat indefinitely.

Our description of the oscillator is an idealization where resistive forces such as friction and air resistance have been neglected. A typical resistive force would be proportional to the angular velocity of the mass. This leads us to consider the following differential equation

$$\ddot{\phi} \quad + \quad \gamma\dot{\phi} + \omega_0^2\phi = 0, \tag{4.10}$$

where γ is a constant related to the strength of the resistance. In order to fully describe the undamped oscillator we needed two linearly independent solutions. To find two such functions for (4.10) we can try $x = e^{\lambda t}$ where λ is a complex number to be determined. Plugging into (4.10) we will find

$$\lambda^2 e^{\lambda t} + \gamma \lambda e^{\lambda t} + \omega_0^2 e^{\lambda t} = 0,$$
$$\Rightarrow \lambda^2 + \gamma \lambda + \omega_0^2 = 0,$$

$$\Rightarrow \lambda = \frac{-\gamma \pm \sqrt{\gamma^2 - 4\omega_0^2}}{2}. \tag{4.11}$$

Let

$$k = \gamma^2 - 4\omega_0^2,$$
$$\kappa = |k|.$$

The behavior of the system depends on the value of k.

Case 1. $k < 0$

$$\lambda = \frac{-\gamma \pm i\sqrt{\kappa}}{2}. \tag{4.12}$$

We can write the general solution:

$$\phi(t) = \alpha e^{-\sigma t + i \Upsilon t} + \beta e^{-\sigma t - i \Upsilon t},$$
$$\text{where}$$
$$\sigma = \frac{\gamma}{2},$$
$$\Upsilon = \frac{\sqrt{\kappa}}{2}.$$

Hence, the general solution may be written

$$\phi(t) = e^{-\sigma t}[\alpha e^{i \Upsilon t} + \beta e^{-i \Upsilon t}], \tag{4.13}$$

or equivalently

$$x(t)^{-\sigma t}[A \cos(\Upsilon t + \eta)]. \tag{4.14}$$

The system will still oscillate but the magnitude of the oscillation will be reduced by the exponentially decaying factor $e^{-\sigma t}$. Figure 4.3 shows a particular example of such motion. In the left panel the plot of ϕ against time. The dashed curves correspond to the bounding curves $\pm e^{-\sigma t}$ and in the left the phase trajectory which spirals toward the point $(0,0)$ we will call such a point an "*attractor.*"

This case, where we still see oscillations, is described as under damped.

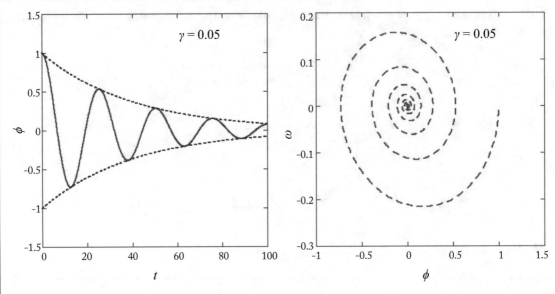

Figure 4.3: Left panel ϕ plotted against t for $\gamma = 0.05$, also shown are the bounding curves, $\pm e^{-0.025t}$; right panel is the phase trajectory.

Case 2. $k > 0$

In this case both

$$\frac{-\gamma - \sqrt{\gamma^2 - 4\omega_0^2}}{2}$$

and

$$\frac{-\gamma + \sqrt{\gamma^2 - 4\omega_0^2}}{2},$$

are negative real numbers and the solution

$$Ae^{\frac{-\gamma + \sqrt{\gamma^2 - 4\omega_0^2}}{2}} + Be^{\frac{-\gamma - \sqrt{\gamma^2 - 4\omega_0^2}}{2}} \tag{4.15}$$

just decays with time and shows no oscillation. Such a solution is said be over damped.

Case 3. $k = 0$

This case which is know as critical damping marks the transition from oscillatory to decaying behavior. Mathematically, it is a little different in that we have only one independent solution $e^{-\lambda t}$. However, it is easy to check that in this special case $te^{-\lambda t}$ is a second solution and indeed the general solution can be written; see for example [16]

$$\phi(t) = Ae^{-\frac{\gamma}{2}t} + Bte^{-\frac{\gamma}{2}t}. \tag{4.16}$$

In this case we see no oscillations.

Lemma 4.1   *Let $x_g(t)$ be the general solution of the homogenous 2nd order linear differential equation*

$$a_2 \frac{d^2 x}{dt^2} + 2a_1 \frac{dx}{dt} + a_0 x = 0,$$

and x_p is any particular solution of the inhomogeneous differential equation

$$a_2 \frac{d^2 x}{dt^2} + 2a_1 \frac{dx}{dt} + a_0 x(t) = f(t)$$

then any other solution $X(t)$ must be of the form

$$X(t) = x_p(t) + x_g(t).$$

Proof.

$$a_2 \frac{d^2 x_p}{dt^2} + 2a_1 \frac{dx_p}{dt} + a_0 x_p(t) = f(t),$$

$$a_2 \frac{d^2 X}{dt^2} + 2a_1 \frac{dX}{dt} + a_0 X(t) = f(t),$$

$$\Rightarrow a_2 \frac{d^2 x_p - X}{dt^2} + 2a_1 \frac{dx_p - X}{dt} + a_0 [x_p(t) - X(t)] = 0.$$

Therefore, $x_p - X$ is a solution of the homogenous problem and as such we must have $x_p - X = x_g(t)$ $x_c(t)$ will contain two constants so we have enough to accommodation the boundary conditions so

$$X(t) = x_g(t) + x_p(t).$$

□

Suppose, now, we wish to solve the differential equation

$$\ddot{\phi} = -\gamma \dot{\phi} - \omega_0^2 \phi + Q \sin(\Omega t). \tag{4.17}$$

In other words, we are applying an external oscillating force to our pendulum. This could be achieved, for example, if the mass is charged and we apply an external varying electric field. We are assuming γ, ω_0^2 and Q are positive real constants. Since the homogenous part of (4.17) is just the damped oscillator this means we know the general solution. So "*all*" that is needed is a particular solution. To this end it is useful to look at the complex generalization of (4.17)

$$\ddot{z} = -\gamma \dot{z} - \omega_0^2 z + Q e^{i\Omega t}. \tag{4.18}$$

Our desired particular solution will be the imaginary part of some z_p: a particular solution of (4.18). The form (4.18) is suggestive of a possible solution of the form

$$z = z_0 e^{i\Omega t}.$$

Plugging this into (4.18) we have

$$
\begin{aligned}
[-\Omega^2 z_0 e^{i\Omega t} + i\Omega \gamma z_0 e^{i\Omega t} + z_0 e^{i\Omega t} \omega_0^2] &= Q e^{i\Omega t}, \\
\Rightarrow z_0[(\omega_0^2 - \Omega^2) + i\Omega \gamma] &= Q, \\
\Rightarrow z_0 &= \frac{Q}{(\omega_0^2 - \Omega^2) + i\gamma\Omega}, \\
\Rightarrow z_0 &= \frac{Q}{(\omega_0^2 - \Omega^2) + i\gamma\Omega} \frac{(\omega_0^2 - \Omega^2) - i\gamma\Omega}{(\omega_0^2 - \Omega^2) - i\gamma\Omega} \\
&= |z_0| e^{i\chi}, \tag{4.19}
\end{aligned}
$$

where

$$
\begin{aligned}
|z_0| &= \frac{Q}{\sqrt{(\omega_0^2 - \Omega^2)^2 + \gamma^2 \Omega^2}}, \\
\cos \chi &= \frac{\omega_0^2 - \Omega^2}{\sqrt{[\omega_0^2 - \Omega^2]^2 + \gamma^2 \Omega^2}}, \\
\sin \chi &= -\frac{\gamma\Omega}{\sqrt{[\omega_0^2 - \Omega^2]^2 + \gamma^2 \Omega^2}}, \\
\chi &= \arctan\left(\frac{\gamma\Omega}{\Omega^2 - \omega_0^2}\right). \tag{4.20}
\end{aligned}
$$

For the under-damped forced oscillator with $k < 0$ we have, making use of (4.13) and Lemma 4.1 we find

$$\phi(t) = \alpha e^{-\sigma t + i\Upsilon t} + \beta e^{-\sigma t - i\Upsilon t} + \frac{Q}{\sqrt{(\omega_0^2 - \Omega^2)^2 + \gamma^2 \Omega^2}} \sin(\Omega t + \chi). \tag{4.21}$$

The first two term on the right-hand side will decay with time, the third term will become dominant. If $\gamma << 1$ and $\omega \approx \omega_0$ the amplitude of this term can be very large. This is phenomenon is known as resonance. In Figure 4.4, we show an example of the damped-driven oscillator with $Q = 1, \Omega = 2.0, \gamma = 0.1$. Initially, the system is seen to initially exhibit damped, irregular motion (transient behavior), but eventually it settles into a periodic motion with the same frequency as the driving force. There is a phase-space attractor in this case as well: a closed loop.

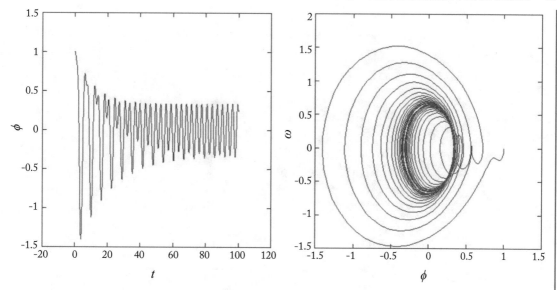

Figure 4.4: Damped-driven pendulum, small oscillations, with $\gamma = 0.1$, $\Omega = 2$, $\omega_0 = 1.0$; left panel time evolution $\phi(t)$, right panel phase trajectory.

4.2.2   DIFFERENCES BETWEEN LINEAR AND NONLINEAR PENDULUM

First let us compare the small angle case, $\gamma = q = 0$, $\phi_0 = 0.1$ radians ($\approx 5.7°$) $\dot{\phi}_0 = 0$.

From the numerical stand point we can just as well use our Runge–Kutta approach for the physical pendulum, even though the underlying equation, (4.6) is nonlinear. As a first example let us consider the simple oscillator, with no damping and no forcing term and consider the difference between the linear and nonlinear cases when we change the initial position of the mass. In Figure 4.5, the mass is released from rest at an angle $\phi = 1$ radian and the linear and nonlinear solutions are nearly identical. If we release the mass from an angle of 60° the nonlinear solution continues to be periodic but out of phase with the linear case; see Figure 4.6. This is not surprising since energy must be conserved, and we have switched off the dissipative force all the initial potential energy must be converted into kinetic so the pendulum will rise until all the kinetic energy has been converted to potential.

Example 4.2   Let us now explore the behavior of the linear and nonlinear oscillator when both are forced and damped. The equation for the linear oscillator is given by (4.17)

$$\ddot{\phi} = -\gamma\dot{\phi} - \omega_0^2\phi + Q\sin(\Omega t),$$

and the nonlinear case by

$$\ddot{\phi} = -\gamma\dot{\phi} - \omega_0^2\sin(\phi) + Q\sin(\Omega t). \tag{4.22}$$

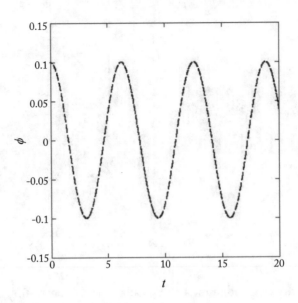

Figure 4.5: Undamped undriven oscillator: blue linear approximation red nonlinear $\gamma = q = 0, \phi_0 = 1$ radian ($\approx 5.7°$) $\dot{\phi}_0 = 0$.

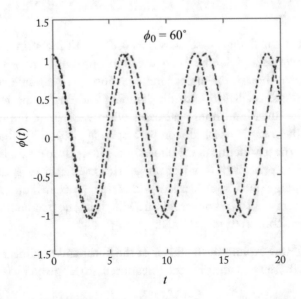

Figure 4.6: Undamped undriven oscillator: blue linear approximation red nonlinear $\gamma = q = 0, \phi_0 = \frac{\pi}{3}$ radian (60°) $\dot{\phi}_0 = 0$.

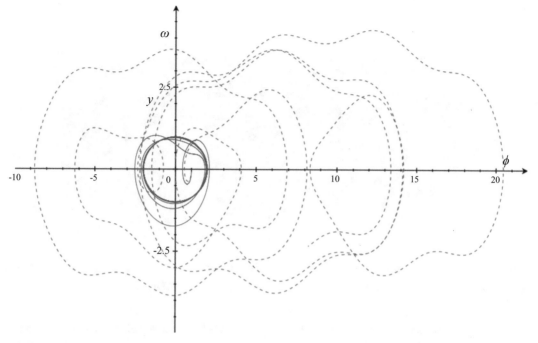

Figure 4.7: Phase trajectories, i.e., $\dot{\phi} = \omega$ against ϕ, for the damped un-driven oscillator with $Q = 2.5, \Omega = 0.5, \gamma = 0.2, \omega_0 = 1$. ϕ is in radians, $\omega = \dot{\phi}$ in radians/s, red dashed nonlinear, blue solid, linear.

I ran a fourth-order Runge–Kutta code for $0 \leq t \leq 100$ and a step size of 0.01, with initial conditions $\phi(0) = 1, \dot{\phi}(0) = 0$ for both the linear and nonlinear cases where I have chosen $Q = 2.5, \Omega = \frac{1}{2}, \omega_0 = 1, \gamma = 0.2$. The results are shown in Figure 4.7.

The linear case settles down to a behavior similar to that we have seen already in Figure 4.4. The phase trajectories for the nonlinear case are much more erratic and do not become less so as we increase the time range. The nonlinear case is very sensitive to the of step and also to the value of Q. These type of sensitivities occur frequently when dealing with nonlinear systems.

4.3 CHAOS

Many mechanical systems exhibit chaotic motion in some regions of their parameter spaces. Essentially, the term *chaotic motion*, or *chaos*, refers to aperiodic motion and sensitivity of the time evolution to the initial conditions. A chaotic system is in practice unpredictable on long time scales, although the motion is in principle deterministic, because minute changes in the initial conditions can lead to large changes in the behavior after some time. Although a chaotic system is unpredictable, its motion is not completely random. In particular, the way the system

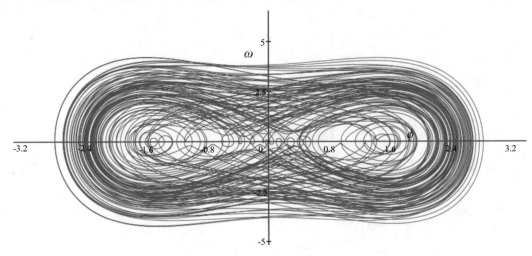

Figure 4.8: Phase trajectory for the Duffing oscillator with $\gamma = 0.1, \alpha = 1, \beta = -1, Q = 2.4$, and initial conditions $\phi_0 = 1.0, \dot{\phi}_0 = 0, 0 \le t \le 500$ calculated with a step size of 0.1.

approaches chaos often exhibits universality, i.e., seemingly different systems make the transition from regular, periodic motion to chaotic motion in very similar ways, often through a series of quantitatively universal period doublings (bifurcations). Our nonlinear pendulum can be shown to exhibit chaotic motion. A discussion of this important topic can be found in [17]. One of the first chaotic systems to be studied was the Duffing oscillator:

$$\ddot{\phi} - \gamma\dot{\phi} + \alpha x + \beta x^3 = Q \sin(\Omega t). \tag{4.23}$$

It occurs in a number of situations, for example in the modeling of a damped-driven elastic pendulum whose spring does not exactly obey Hooke's Law. In Figure 4.8, the phase trajectory for a Duffing oscillator with $\gamma = 0.1, \alpha = 1, \beta = -1$, and $Q = 2.4$ are shown. For this case the system is chaotic.

CHAPTER 5

Numerical Linear Algebra

A vast number of problems in computational physics can be reduced to the solution of systems of linear equations and the related problem of finding solutions to the matrix eigenvalue equation:

$$Ax = \lambda x, \tag{5.1}$$

where A is an $N \times N$ matrix, x is a $N \times 1$ matrix (column vector), and λ is a number. In this chapter, I will develop some of the basic numerical methods for dealing with large systems of equations and the eigenvalue problem (5.1). The development of the numerical methods I will present below assumes a basic knowledge of linear algebra. A brief summary of the major results I will need is presented in Appendix A.

5.1   SYSTEM OF LINEAR EQUATIONS

Suppose we need to find the unknowns $\{x_i\}_{i=1}^{N}$ that solve a system of linear equations of the form:

$$
\begin{aligned}
a_{11}x_1 + a_{21}x_2 + \cdots + a_{1N}x_N &= b_1, \\
a_{21}x_1 + a_{22}x_2 + \cdots + a_{2N}x_N &= b_2, \\
&\vdots \\
a_{N1}x_1 + a_{2N}x_2 + \cdots + a_{NN}x_N &= b_N,
\end{aligned}
\tag{5.2}
$$

where the numbers a_{ij}, b_j are known. We can write (5.2) as a matrix equation $Ax = b$:

$$
\begin{bmatrix}
a_{11} & a_{21} & \cdots & a_{1N} \\
a_{21} & a_{22} & \cdots & a_{2N} \\
\vdots & \vdots & \ddots & \vdots \\
a_{N1} & a_{N2} & \cdots & a_{NN}
\end{bmatrix}
\begin{pmatrix}
x_1 \\ x_2 \\ \vdots \\ x_N
\end{pmatrix}
=
\begin{pmatrix}
b_1 \\ b_2 \\ \vdots \\ b_N
\end{pmatrix}.
\tag{5.3}
$$

If the $N \times N$ matrix A is non singular our desired solution is

$$x = A^{-1}b.$$

Now, see Appendix A,

$$A^{-1} = \frac{1}{|A|}C^T, \tag{5.4}$$

$|A|$ is the determinant of A and C is the cofactor matrix corresponding to A. To calculate the inverse of a $N \times N$ matrix using the cofactor method would involve something of the order of $N!$ multiplications. For $N = 20$, this would mean approximately 2×10^{18} multiplications. Even with today's fast machines it would take a long time to solve a system of equations this way. We need a better computational approach.

There are some particular cases where the solution of (5.3) is particularly easy.

- If A is diagonal

$$
A = \begin{bmatrix} a_{11} & 0 & \cdots & 0 \\ 0 & a_{22} & \cdots & 0 \\ \vdots & \vdots & \ddots & 0 \\ 0 & 0 & \cdots & a_{NN} \end{bmatrix},
$$

then the linear equations are uncoupled and

$$
x_i = \frac{b_i}{a_{ii}} \quad i = 1, \ldots, N. \tag{5.5}
$$

- If A is an upper triangular matrix:

$$
A = \begin{bmatrix} a_{11} & a_{12} & \cdots & a_{1N} \\ 0 & a_{22} & \cdots & a_{2N} \\ 0 & \vdots & \ddots & \vdots \\ 0 & 0 & \cdots & a_{NN} \end{bmatrix},
$$

where all the elements below the main diagonal are zero, $a_{ij} = 0, \forall i > j$, then

$$
\begin{bmatrix} a_{11} & a_{12} & \cdots & a_{1N} \\ 0 & a_{22} & \cdots & a_{2N} \\ 0 & \vdots & \ddots & \vdots \\ 0 & 0 & \cdots & a_{NN} \end{bmatrix} \begin{pmatrix} x_1 \\ x_2 \\ \vdots \\ x_N \end{pmatrix} = \begin{pmatrix} b_1 \\ b_2 \\ \vdots \\ b_N \end{pmatrix} \tag{5.6}
$$

admits a "*backward substitution*" solution

$$
\begin{aligned}
a_{NN} x_N &= b_N, \\
a_{N-1N-1} x_{N-1} + a_{N-1N} x_N &= b_{N-1}, \\
\vdots &= \vdots \\
\sum_{j=1}^{N} a_{1i} x_i &= b_1.
\end{aligned} \tag{5.7}
$$

- In much the same way we can solve the linear equations when A is a lower triangular matrix using a "*forward substitution*" solution.

My basic strategy here is to look for ways to relate our matrix A to one or other of these simpler forms.

5.2   LU FACTORIZATION

Suppose A can be decomposed into the product of a lower-triangular matrix L and an upper-triangular matrix U. The entire solution algorithm for $Ax = b$ can be described in three steps

(i) Decompose $A = LU$.

(ii) Solve : $Ly = b$.

(iii) Solve $Ux = y$.

Example 5.1   Consider the decomposition of the matrix

$$A = \begin{pmatrix} 3 & 1 \\ 4 & 2 \end{pmatrix}.$$

We can require

$$\begin{pmatrix} 3 & 1 \\ 4 & 2 \end{pmatrix} = \begin{pmatrix} l_{11} & 0 \\ l_{21} & l_{22} \end{pmatrix} \begin{pmatrix} u_{11} & u_{12} \\ 0 & u_{22} \end{pmatrix},$$

$$\begin{aligned} \Rightarrow 3 &= l_{11}u_{11}, \\ 1 &= l_{11}u_{12}, \\ 4 &= l_{21}u_{11}, \\ 2 &= l_{21}u_{12} + l_{22}u_{22}. \end{aligned}$$

We have only four equations and we need six unknowns, however we only require a decomposition. So just take $l_{11} = 1, l_{22} = 1$. Then,

$$\begin{pmatrix} 4 & 3 \\ 6 & 3 \end{pmatrix} = \begin{pmatrix} 1 & 0 \\ l_{21} & 1 \end{pmatrix} \begin{pmatrix} u_{11} & u_{12} \\ 0 & u_{22} \end{pmatrix}.$$

$$\begin{aligned} \Rightarrow 3 &= u_{11}, \\ 1 &= u_{12}, \\ 4 &= l_{21}u_{11}, \\ 2 &= l_{21}u_{12} + u_{22}. \end{aligned}$$

$$\Rightarrow \begin{pmatrix} 4 & 3 \\ 6 & 3 \end{pmatrix} = \begin{pmatrix} 1 & 0 \\ \frac{4}{3} & 1 \end{pmatrix} \begin{pmatrix} 3 & 1 \\ 0 & \frac{2}{3} \end{pmatrix}.$$

For a further discussion of the LU approach, see [15].

5.3   QR FACTORIZATION

The QR approach is an alternative decomposition where a non singular matrix A is "decomposed" into the product of a unitary matrix and an upper triangular matrix.

Theorem 5.2   *Suppose that Q is an $N \times N$ non singular matrix acting on an N dimensional complex space $\mathbb{V}$.*

   Q is unitary iff the column (row) vectors of Q form an orthonormal basis for $\mathbb{V}$.

Proof. If we write Q in the form:

$$
Q = \begin{bmatrix}
c_{11} & c_{12} & \cdots & c_{1N} \\
c_{21} & c_{22} & \cdots & c_{2N} \\
\vdots & \vdots & \ddots & \vdots \\
c_{N1} & c_{N2} & \cdots & c_{NN}
\end{bmatrix},
$$

then we can regard Q as being made up of N vectors

$$
c_i = \begin{bmatrix}
c_{1i} \\
c_{2i} \\
\vdots \\
c_{Ni}
\end{bmatrix},
$$

$$
Q = (c_1 c_2 \cdots c_N) \tag{5.8}
$$

and

$$
\langle c_i | c_j \rangle = \sum_{k=1}^{N} \bar{c}_{ki} c_{kj},
$$

$$
Q^\dagger Q = \begin{bmatrix}
\langle c_1 | c_1 \rangle & \langle c_1 | c_2 \rangle, & \cdots, & \langle c_1 | c_N \rangle \\
\langle c_1 | c_2 \rangle & \langle c_2 | c_2 \rangle, & \cdots & \langle c_2 | c_N \rangle \\
\vdots & \vdots & \ddots & \vdots \\
\langle c_1 | c_N \rangle & \langle c_2 | c_N \rangle, & \cdots & \langle c_N | c_N \rangle
\end{bmatrix}.
$$

Thus, Q is unitary iff $\langle c_i | c_j \rangle = \delta_{ij}$.

The proof using rows as vectors is almost identical. $\square$

Suppose $A \in \mathbb{R}^{N \times N}$ is a non singular matrix then we know its determinant is non zero and the vectors corresponding to the columns are linearly independent. Just as in (5.8) I can write

$$
A = [a_1 a_2 \ldots a_N].
$$

My plan is to use the Grahm–Schmidt process to create an orthomormal set from the vectors a_i. We may write using (A.6)

$$
\begin{aligned}
e_1' &= a_1, \\
e_1 &= \frac{e_1'}{\|e_1'\|}, \\
e_2' &= a_2 - \langle a_2 | e_1 \rangle e_1, \\
e_2 &= \frac{e_2'}{\|e_2'\|}, \\
&\vdots
\end{aligned}
\tag{5.9}
$$

The vector a_i is a member of the subspace spanned by $\{e_k\}_{k=1}^i$ and thus orthogonal to the unit vectors $\{e_k\}_{k>i}$. We can expand

$$
\begin{aligned}
a_i &= \sum_{k=1}^{N} r_{ki} e_k \\
r_{ki} &= \langle e_k | a_i \rangle.
\end{aligned}
\tag{5.10}
$$

Notice $r_{ki} = 0$ if $k > i$. Further,

$$
\begin{aligned}
r_{ii} &= \langle e_i | e_i' \rangle \\
&= \langle e_i' | e_i' \rangle \geq 0.
\end{aligned}
$$

Considering components, the (li)th term from our original matrix A, this is given by the lth element of a_i can, therefore, be written in terms of the lth elements of e_k

$$
\begin{aligned}
a_{li} &= \sum_{k=1}^{N} r_{ki} e_{lk} \\
&= \sum_{k=1}^{N} e_{lk} r_{ki} \\
A &= QR,
\end{aligned}
$$

where by construction the matrix

$$
Q = [e_1 e_2 \dots e_N]
$$

satisfy the unitary condition in Theorem 5.2 and the matrix, R has only zero elements below the diagonal, i.e., it is upper triangular.

Example 5.3   Consider the matrix

$$
A = \begin{pmatrix} 3 & 2 \\ 1 & 2 \end{pmatrix},
$$

writing the columns as vectors

$$A = [a_1 a_2],$$

where

$$a_1 = \begin{pmatrix} 3 \\ 1 \end{pmatrix},$$

$$a_2 = \begin{pmatrix} 2 \\ 2 \end{pmatrix}.$$

Suppose $\Omega = \text{Span}\{a_1, a_2\}$. Applying the Grahm–Schmidt process to a_1, a_2 you will find the two orthonormal vectors

$$e_1 = \frac{1}{\sqrt{10}} \begin{pmatrix} 3 \\ 1 \end{pmatrix}, \quad e_2 = \frac{1}{\sqrt{10}} \begin{pmatrix} -1 \\ 3 \end{pmatrix},$$

which also span Ω. Define

$$Q = \frac{1}{\sqrt{10}} \begin{pmatrix} 3 & -1 \\ 1 & 3 \end{pmatrix},$$

$$\Rightarrow Q^\dagger = \frac{1}{\sqrt{10}} \begin{pmatrix} 3 & 1 \\ -1 & 3 \end{pmatrix}.$$

$$Q^\dagger Q = I$$

$$= \begin{pmatrix} 1 & 0 \\ 0 & 1 \end{pmatrix}.$$

$$A = QR,$$
$$Q^\dagger A = R$$
$$= \frac{1}{\sqrt{10}} \begin{pmatrix} 3 & 1 \\ -1 & 3 \end{pmatrix} \begin{pmatrix} 3 & 2 \\ 1 & 2 \end{pmatrix},$$
$$= \frac{1}{\sqrt{10}} \begin{pmatrix} 10 & 8 \\ 0 & 4 \end{pmatrix}.$$

As we have seen every non singular matrix A can be decomposed into a product QR, where Q is unitary and R is upper triangular. Let us now apply this observation to some of the characteristic problems of linear algebra.

5.3.1   SYSTEMS OF LINEAR EQUATIONS

Suppose A is a non-singular $N \times N$ matrix and we want to solve

$$A x = b.$$

We can proceed as follows.

- First factorize A

$$A x = Q R x = b.$$

- Then introduces a vector y

$$
\begin{aligned}
y &\equiv R x, \\
\Rightarrow Q y &= b, \\
\Rightarrow y &= Q^\dagger b.
\end{aligned}
$$

- Solve the triangular system $R x = y$ by back substitution.

5.3.2   EIGENVALUES

Our next task is to find the eigenvalues of A a non singular $N \times N$ matrix. We first note that if R is an upper triangular matrix, its eigenvalues are given by the solution of

$$
|R - \lambda I| = \begin{vmatrix}
a_{11} - \lambda & a_{12} & \cdots & a_{1N} \\
0 & a_{22} - \lambda & \cdots & a_{2N} \\
0 & & \ddots & \vdots \\
0 & 0 & \cdots & a_{NN} - \lambda
\end{vmatrix} = 0,
$$

$$\Rightarrow |R - \lambda I| = \prod_{i=1}^{N} (a_{ii} - \lambda) = 0.$$

Now we know, see Lemma A.29, that if A is an $N \times N$ matrix and U is a unitary matrix then if $B = UAU^\dagger$ then B and A have the same eigenvalues. If we can find a matrix B which is upper triangular such that

$$A = UBU^\dagger \tag{5.11}$$

and U is unitary then the eigenvalues of A are just the diagonal elements of B. We will call a transformation of the kind (5.11) with U unitary a "*similarity transformation.*"

We can find a QR decomposition of any non singular matrix A, i.e.,

$$
\begin{aligned}
A &= QR, \\
Q^\dagger A Q &= RQ,
\end{aligned}
\tag{5.12}
$$

is a similarity transformation of A and RQ has the same eigenvalues as $A = QR$. and if this is upper diagonal the problem is solved. Even if RQ is not in upper triangular form we do have a way forward.

The QR Algorithm

Suppose A is a real symmetric matrix for which we want to find the eigenvalues. Define $A^0 = A$, then define a sequence of matrices starting with $k = 0$ by computing the QR decomposition to find $Q^{(k)}$ and $R^{(k)}$ then define $A^{(k+1)} = R^k Q^k$. Now, it can be shown that eventually $A^{(k)}$ converges to an upper triangular matrix [18]. But

$$
\begin{aligned}
A^{(k)} &= R^{(k-1)} Q^{(k-1)}, \\
&= (Q^{(k-1)})^\dagger Q^{(k-1)} R^{(k-1)} Q^{(k-1)}, \\
&= (Q^{(k-1)})^\dagger A^{(k-1)} Q^{(k-1)},
\end{aligned}
$$

so all the matrices $A^{(j)}$ are connected by similarity transformations and therefore share the same eigenvalues $A^{(0)}$ is just A and $A^{(k)}$ is upper triangular so we can read off the eigenvalues.

The advantages of the of the QR algorithm are that it

- gives all the eigenvalues,

- is stable.

It is incorporated in the major software packages such as LAPACK [2].

5.3.3   LINEAR LEAST SQUARES

The QR factorization is not restricted to $N \times N$ matrices.

Example 5.4   Consider the matrix:

$$
A = \begin{bmatrix} 1 & 2 \\ 1 & 2 \\ 0 & 3 \end{bmatrix},
$$

write the columns as the vectors

$$
r_1 = \begin{pmatrix} 1 \\ 1 \\ 0 \end{pmatrix},
$$

$$
r_2 = \begin{pmatrix} 2 \\ 2 \\ 3 \end{pmatrix}.
$$

Suppose $\Omega = \text{Span}\{r_1, r_2\}$. Using the Grahm–Schmidt process it is easy to see that the vectors e_1, e_2

$$
e_1 = \frac{1}{\sqrt{2}} \begin{pmatrix} 1 \\ 1 \\ 0 \end{pmatrix},
$$

$$e_2 = \begin{pmatrix} 0 \\ 0 \\ 1 \end{pmatrix},$$

are orthonormal. Define

$$Q = \begin{pmatrix} \frac{1}{\sqrt{2}} & 0 \\ \frac{1}{\sqrt{2}} & 0 \\ 0 & 1 \end{pmatrix},$$

$$Q^T = \begin{pmatrix} \frac{1}{\sqrt{2}} & \frac{1}{\sqrt{2}} & 0 \\ 0 & 0 & 1 \end{pmatrix},$$

$$\Rightarrow Q^T = I = \begin{pmatrix} 1 & 0 \\ 0 & 1 \end{pmatrix}.$$

Define

$$
\begin{aligned}
R &= Q^\dagger A \\
&= \begin{pmatrix} \frac{1}{\sqrt{2}} & \frac{1}{\sqrt{2}} & 0 \\ 0 & 0 & 1 \end{pmatrix} \begin{bmatrix} 1 & 2 \\ 1 & 2 \\ 0 & 3 \end{bmatrix} \\
&= \begin{pmatrix} \sqrt{2} & 2\sqrt{2} \\ 0 & 3 \end{pmatrix}.
\end{aligned}
$$

Note that

- Q is a 3×2 matrix and $Q^\dagger$ is a 2×3 matrix,
- $Q^\dagger Q = I_2$ but $Q Q^\dagger \neq I_3$,
- since Q is real $Q^\dagger = Q^T$.

This example is a particular case of the following theorem; see [15].

Theorem 5.5   *Suppose $A \in \mathbb{R}^{M \times N}$ where $M \geq N$ then A can be written in the form*

$$A = QR,$$

where R is an upper triangular $N \times N$ matrix and Q is an $M \times N$ matrix which satisfies

$$Q^T Q = I_N,$$

where I_N is the $N \times N$ identity matrix. If $rank(A) = N$ then R is non singular.

Again notice our matrix Q is real so $Q^\dagger = Q^T$.

Table 5.1: The right-hand side column displays measured values at several times t

t	$E(t)$
1.0	1.0
2.0	1.5
3.0	3.0
4.0	6

Linear Least Squares

Suppose that we are given a vector $b \in \mathbb{R}^M$ and a $M \times N$ matrix A then the matrix equation

$$A x = b \tag{5.13}$$

has, in general, no solutions if $M > N$ but it has an infinity of solutions if $M < N$, there are more equations than unknowns and the system is said be "over determined."

Example 5.6   Table 5.1 shows measured function values of the quantity E at four times. We want to find the "best fit" to the data which can contain some degree of experimental error.

More generally, suppose we have M experimental observations which gives us a set $\{x_i, b_i\}_{i=1}^M$ Suppose E is the desired experimental quantity and we want to model E with a linear combination of N functions $\phi_j(t)$.

We can write

$$E(x) = \sum_{j=1}^{N} c_j \phi_j(x),$$
$$E(x_i) \approx b_i, \; i = 1, \ldots, M, \tag{5.14}$$

which we can put in matrix form as

$$A x = \begin{bmatrix} \phi_1(x_1) & \cdots & \phi_N(x_1) \\ \vdots & \ddots & \vdots \\ \phi_1(x_N) & \cdots & \phi_N(x_N) \\ \vdots & \ddots & \vdots \\ \phi_1(x_M) & \cdots & \phi_N(x_M) \end{bmatrix} \begin{bmatrix} c_1 \\ \vdots \\ c_N \end{bmatrix} = \begin{bmatrix} E(x_1) \\ \vdots \\ E(x_N) \\ \vdots \\ E(x_M) \end{bmatrix} \approx \begin{bmatrix} b_1 \\ \vdots \\ b_N \\ \vdots \\ b_M \end{bmatrix} = b. \tag{5.15}$$

I still have not said precisely what I mean by a "*best fit*," the way we will explore this idea here is to minimize the sum of squares of the deviation i.e., we want to find the vector c which minimizes

$$\sum_{k=1}^{M} (E(x_i) - b_i)^2 = ||Ac - b||^2. \tag{5.16}$$

We can use our *QR* decomposition to find the solution to the least squared problem.

Theorem 5.7   *Suppose that $A \in \mathbb{R}^{M \times N}$ with $M \geq N$ and rank$(A) = N$. Then there exists a unique least squared solution to the system of equations*

$$Ac = b$$

which minimizes

$$||Ac - b||.$$

Further, if we decompose

$$A = QR.$$

Then the vector c defined by

$$Rc = Q^T b \tag{5.17}$$

is that unique solution.

Proof. [15]. □

Returning to Example 5.6. Using the **QR** approach I will look for the "*best fit*" in the least square sense assuming the data is best represented by a

(i) straight line,

(ii) a quadratic function

$$p(t) = c_1 + c_2 t + c_3 t^2.$$

Example 5.8   First we will look for a linear solution $c_0 + c_1 t \approx E(t)$.
In matrix form,

$$Ac = b,$$
$$\begin{bmatrix} 1 & 1 \\ 1 & 2 \\ 1 & 3 \\ 1 & 4 \end{bmatrix} \begin{pmatrix} c_0 \\ c_1 \end{pmatrix} = \begin{bmatrix} 1 \\ 1.5 \\ 3 \\ 6 \end{bmatrix}.$$

Now write A as two column vectors:

$$a_1 = \begin{pmatrix} 1 \\ 1 \\ 1 \\ 1 \end{pmatrix}, \quad a_2 = \begin{pmatrix} 1 \\ 2 \\ 3 \\ 4 \end{pmatrix}.$$

Applying Grahm–Schmidt to these vectors we get:

$$e_1 = \frac{1}{2}\begin{pmatrix} 1 \\ 1 \\ 1 \\ 1 \end{pmatrix}, \quad e_2 = \frac{1}{\sqrt{5}}\begin{pmatrix} -1.5 \\ -0.5 \\ 0.5 \\ 1.5 \end{pmatrix}.$$

Now construct Q

$$Q = \begin{pmatrix} \frac{1}{2} & -\frac{1.5}{\sqrt{5}} \\ \frac{1}{2} & -\frac{0.5}{\sqrt{5}} \\ \frac{1}{2} & \frac{0.5}{\sqrt{5}} \\ \frac{1}{2} & \frac{1.5}{\sqrt{5}} \end{pmatrix}.$$

Then R will be given by

$$\begin{aligned}
R &= Q^T A, \\
&= \begin{pmatrix} \frac{1}{2} & \frac{1}{2} & \frac{1}{2} & \frac{1}{2} \\ -\frac{3}{2\sqrt{5}} & -\frac{1}{2\sqrt{5}} & -\frac{1}{2\sqrt{5}} & \frac{3}{2\sqrt{5}} \end{pmatrix} \begin{bmatrix} 1 & 1 \\ 1 & 2 \\ 1 & 3 \\ 1 & 4 \end{bmatrix} \\
&= \begin{bmatrix} 2 & 5 \\ 0 & \sqrt{5} \end{bmatrix}.
\end{aligned}$$

Now we can deduce values for our coefficients:

$$\begin{aligned}
Ac &= b, \\
\Rightarrow QRc &= b, \\
\Rightarrow Rc &= Q^T b,
\end{aligned}$$

$$\begin{bmatrix} 2 & 5 \\ 0 & \sqrt{5} \end{bmatrix}\begin{bmatrix} c_0 \\ c_1 \end{bmatrix} = \begin{pmatrix} \frac{1}{2} & \frac{1}{2} & \frac{1}{2} & \frac{1}{2} \\ -\frac{3}{2\sqrt{5}} & -\frac{1}{2\sqrt{5}} & -\frac{1}{2\sqrt{5}} & \frac{3}{2\sqrt{5}} \end{pmatrix}\begin{bmatrix} 1 \\ 1.5 \\ 3 \\ 6 \end{bmatrix},$$

$$\begin{pmatrix} 2c_0 + 5c_1 \\ \sqrt{5}c_1 \end{pmatrix} = \begin{pmatrix} 5.75 \\ \frac{8.25}{\sqrt{5}} \end{pmatrix}.$$

Hence, the best linear approximation is

$$y(t) = c_0 + c_1 t = -1.25 + 1.65t. \tag{5.18}$$

Let us now consider a quadratic fit. We are looking for a solution

$$c_0 + c_1 t + c_2 t^2 \approx E(t).$$

We will proceed much as before. We can write the problem in matrix form

$$Ac = b,$$

$$\begin{pmatrix} 1 & 1 & 1 \\ 1 & 2 & 4 \\ 1 & 3 & 9 \\ 1 & 4 & 16 \end{pmatrix} \begin{pmatrix} c_0 \\ c_1 \\ c_2 \end{pmatrix} = \begin{pmatrix} 1 \\ 1.5 \\ 3 \\ 6 \end{pmatrix}.$$

Applying Grahm–Schmidt we find our desired Q to be

$$Q = \frac{1}{2} \begin{pmatrix} 1 & -\frac{3}{\sqrt{5}} & 1 \\ 1 & -\frac{1}{\sqrt{5}} & -1 \\ 1 & \frac{1}{\sqrt{5}} & -1 \\ 1 & \frac{3}{\sqrt{5}} & 1 \end{pmatrix}.$$

Now since Q and A are known all we have to do is transpose Q and multiply by A to find R:

$$R = Q^T A,$$

$$= \frac{1}{2} \begin{pmatrix} 1 & 1 & 1 & 1 \\ -\frac{3}{\sqrt{5}} & -\frac{1}{\sqrt{5}} & \frac{1}{\sqrt{5}} & \frac{3}{\sqrt{5}} \\ 1 & -1 & -1 & 1 \end{pmatrix} \begin{pmatrix} 1 & 1 & 1 \\ 1 & 2 & 4 \\ 1 & 3 & 9 \\ 1 & 4 & 16 \end{pmatrix},$$

$$= \begin{pmatrix} 2 & 5 & 15 \\ 0 & \sqrt{5} & 5\sqrt{5} \\ 0 & 0 & 2 \end{pmatrix}.$$

Then,

$$Rc = Q^T b,$$

$$\begin{pmatrix} 2 & 5 & 15 \\ 0 & \sqrt{5} & 5\sqrt{5} \\ 0 & 0 & 2 \end{pmatrix} \begin{pmatrix} c_0 \\ c_1 \\ c_2 \end{pmatrix} = \frac{1}{2} \begin{pmatrix} 1 & 1 & 1 & 1 \\ -\frac{3}{\sqrt{5}} & -\frac{1}{\sqrt{5}} & \frac{1}{\sqrt{5}} & \frac{3}{\sqrt{5}} \\ 1 & -1 & -1 & 1 \end{pmatrix} \begin{pmatrix} 1 \\ 1.5 \\ 3 \\ 6 \end{pmatrix},$$

$$\begin{pmatrix} 2c_0 + 5c_1 + 15c_2 \\ \sqrt{5}c_1 + 5\sqrt{5}c_2 \\ 2c_2 \end{pmatrix} = \begin{pmatrix} 5.75 \\ \frac{8.25}{\sqrt{5}} \\ 1.25 \end{pmatrix}.$$

$c_2 = 0.625, c_1 = -1.475, c_0 = 1.875.$

So the quadratic least squared fit is:

$$1.875 - 1.475t + 0.625t^2. \tag{5.19}$$

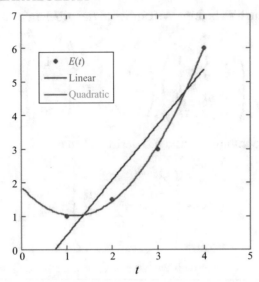

Figure 5.1: Linear least squared and quadratic least squared fits to the data in Table 5.1.

In Figure 5.1 I show a comparison between the linear, (5.18), and quadratic, (5.19), fits.

I cannot end this chapter without adding one word of warning. A straightforward computer implementation of the Grahm–Schmidt process as given in (5.9) can very easily run into significant round off errors. Fortunately, there are some clever ways to avoid these problems [19] leading to stable and efficient codes.

CHAPTER 6

Polynomial Approximations

6.1 INTERPOLATION

Interpolation is the problem of fitting a smooth curve through a given set of points, generally as the graph of a function. It is useful in data analysis (interpolation is a form of regression) and in numerical analysis. It is one of those important recurring concepts in applied mathematics.

Definition 6.1 Given $N + 1$ points $x_j \in \mathbb{R}, 0 \le j \le N$ and sample values

$$y_j = f(x_j)$$

of a function at these points then the polynomial interpolation problem consists in finding a polynomial $p_n(x)$ of degree N which reproduces these values

$$y_j = p_n(x_j).$$

Example 6.2 Linear interpolation between (x_1, y_1) and (x_2, y_2).

Let

$$
\begin{aligned}
L_1(x) &= \frac{x - x_2}{x_1 - x_2}, \\
L_2(x) &= \frac{x - x_1}{x_2 - x_1}, \\
p(x) &= y_1 L_1(x) + y_2 L_2(x), \\
&= y_1 \left[\frac{x - x_2}{x_1 - x_2} \right] + y_2 \left[\frac{x - x_1}{x_2 - x_1} \right], \\
&= \frac{1}{x_1 - x_2} [-y_1 x_2 + y_2 x_1 + x(y_1 - y_2)], \\
&= \frac{y_1 - y_2}{x_1 - x_2} x + \frac{-y_1 x_2 + y_2 x_1}{x_1 - x_2},
\end{aligned}
$$

then

$$
\begin{aligned}
p(x_1) &= y_1, \\
p(x_2) &= y_2.
\end{aligned}
$$

A degree N polynomial can be written as

$$p_N(x) = \sum_0^N a_n x^n.$$

For interpolation the number of degrees of freedom ($N + 1$ coefficients) in the polynomial matches the number of points where the function should fit. If the degree of the polynomial is strictly less than N we cannot in general pass it through all the points (x_i, y_i). Let us look for a direct solution:

$$\sum_{n=0}^N a_n x_j^n = y_j, \quad j = 0, \ldots, N. \tag{6.1}$$

In these $N + 1$ equations the unknowns are the coefficients $a_0, a_1, \ldots, a_N$. In other words, this is a linear system given by the matrix equation

$$Va = y,$$

$$
\begin{bmatrix}
1 & x_0 & \cdots & x_0^N \\
1 & x_1 & \cdots & x_1^N \\
\vdots & \vdots & \ddots & \vdots \\
1 & x_N & \cdots & x_N^N
\end{bmatrix}
\begin{pmatrix}
a_0 \\
a_1 \\
\vdots \\
a_n
\end{pmatrix}
=
\begin{pmatrix}
y_0 \\
y_1 \\
\vdots \\
y_n
\end{pmatrix}. \tag{6.2}
$$

V is known as the "*Vandermonde matrix*," it has coefficients

$$V_{jn} = x_j^n.$$

If we know how to effectively invert V then we can find a as

$$a = V^{-1} y.$$

I will return to this relatively shortly but for the time being I will develop an analytic approach. Let $\mathbb{P}_N$ denote the set of all real valued polynomials of degree $\leq N$.

Lemma 6.3   *Let $\{x_j\}_{j=0}^N$ be a set of distinct numbers then there exists polynomials $L_k(x) \in \mathbb{P}_N$ such that*

$$L_k(x_j) = \delta_{jk}.$$

Proof. For a given k, define

$$L_k(x) = \frac{\prod_{\substack{j=0 \\ j \neq k}}^N (x - x_j)}{\prod_{\substack{j=0 \\ j \neq k}}^N (x_k - x_j)},$$

$$\Rightarrow L_k(x_j) = \delta_{jk}.$$

□

The polynomials $L_k(x)$ are known as the *"Lagrange elementary polynomials."*

Theorem 6.4   (Lagrange interpolation theorem).   *Let $\{x_j\}_{j=0}^N$ be a collection of disjoint real numbers and $\{y_j\}_{j=0}^N$ be a collection of real numbers. Then there exits a unique $p_n \in \mathbb{P}_N$ s.t.*

$$p_N(x_i) = y_i.$$

Proof. Define

$$p_N(x) = \sum_{k=0}^N y_k L_k(x),$$

where $L_k(x)$ are the Lagrange elementary polynomials. Now from Lemma 6.3 we have

$$p_N(x_i) = \sum_{k=0}^N y_k L_k(x_i) = \sum_{k=0}^N y_k \delta_{ik} = y_i.$$

It remains to show that $p_N(x)$ is unique.

Assume that there exist a second polynomial of order N, $q_N(x)$ such that

$$p_N(x_i) = q_N(x_i) = y_i;$$

then,

$$r_N(x) = p_N(x) - q_N(x)$$

is a polynomial of degree N and has roots at each of the $N + 1$ points $x_0, x_1, \ldots, x_N$. However, the fundamental theorem of algebra tells us that a non-zero polynomial of degree N can only have at most N real roots it follows that r_N must be the zero polynomial. So $p_N = q_N$.   □

Definition 6.5   If f is a function defined on an interval containing the points $\{x_i\}_{i=0}^N$ then

$$P_N(x) = \sum_{k=0}^N f(x_k) L_k(x)$$

is called the Lagrange interpolation polynomial of f.

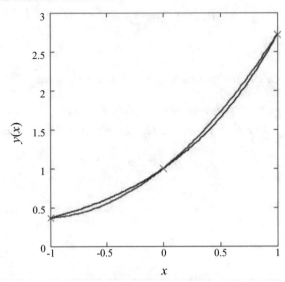

Figure 6.1: Polynomial fit to $\exp(x)$ function evaluated at the points $x = -1, 0, 1$.

Example 6.6 $f(x) = e^x$ interpolated by a parabola with fixed values at $x_0 = -1, x_1 = 0, x_2 = 1$.

$$
\begin{aligned}
L_0(x) &= \frac{(x - x_1)(x - x_2)}{(x_0 - x_1)(x_0 - x_2)}, = \frac{x(x - 1)}{(-1)(-1 - 1)}, = \frac{x^2 - x}{2}. \\
L_1(x) &= \frac{(x - x_0)(x - x_2)}{(x_1 - x_0)(x_1 - x_2)}, = \frac{x + 1(x - 1)}{(1)(-1)}, = 1 - x^2. \\
L_2(x) &= \frac{(x - x_0)(x - x_1)}{(x_2 - x_0)(x_2 - x_2)}. = \frac{x + 1(x)}{(1 + 1)(1)}, = \frac{x^2 + 1}{2}, \\
\Rightarrow p_2(x) &= e^{-1}L_0(x) + e^0 L_1(x) + e^1 L_2(x) = e^{-1}\left(\frac{x^2 - x}{2}\right) + 1 - x^2 + e^1\left(\frac{x^2 + x}{2}\right), \\
&= x^2\left(\frac{e^{-1} + e^1}{2} - 1\right) + x\left(\frac{e - e^{-1}}{2}\right) = x^2(\cosh^2(1) - 1) \\
&+ x\sinh(x) + 1 \approx 1 + 1.1752x + 0.5431x^2.
\end{aligned}
$$

6.1.1   ERROR ESTIMATION

Theorem 6.7   *Let f be a $N + 1$ continuously differentiable function on an interval $[a, b]$ and let $\{x_j | j = 0, \ldots, N\}$ be a set of distinct numbers in $[a, b]$. If $p_N(x)$ is the Langrange interpolation of f*

using $\{x_j \mid j = 0, \ldots, N\}$ then for every $x \in [a,b]$ there exists $\xi(x) \in [a,b]$ s.t.

$$f(x) - p_N(x) = \frac{f^{N+1}(\xi(x))}{(N+1)!}\pi_{N+1}(x),$$

where

$$\pi_{N+1}(x) = \prod_{j=0}^{N}(x - x_j).$$

Proof. [15]. $\square$

The interpolation error:

- depends on the smoothness of f via the high order derivative $f^{(N+1)}$,

- has a factor of $1/(N+1)!$ which decays fast as N becomes large,

- is directly proportional to $\pi_{N+1}(x)$, which means that it will be zero at the points x_j and at its best in their vicinity.

Can we always expect convergence of the polynomial interpolant as $N \to \infty$? The answer is: No!

There are examples of very smooth (analytic) functions for which polynomial interpolation diverges, particularly so near the boundaries of the interpolation interval.

Example 6.8   *Runge function.*
Consider the function

$$f^{runge}(x) = \frac{1}{1 + 25x^2}.$$

Runge found that if it is interpolated at n equidistant points x_i between -1 and 1 the interpolation error increases when the degree of the polynomial is increased; see Figure 6.2.

6.2   ORTHOGONAL POLYNOMIALS

6.2.1   LEGENDRE EQUATION

Legendre polynomials turn up in the solution of a lot of physical problems, in electromagnetism and quantum mechanics [11]. They originate as the solution of the Legendre differential equation

$$(1 - x^2)\frac{d^2y(x)}{dx^2} - 2x\frac{dy(x)}{dx} + l(l+1)y(x) = 0, \tag{6.3}$$

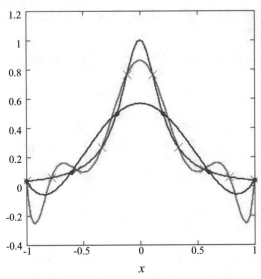

Figure 6.2: Runge function, red line, 5th order interpolation (six equally spaced points) blue line, 9th order interpolation (10 equally spaced points), green line.

where l is a constant. In this section we will start from (6.3) and look for a power series solution

$$\frac{dy}{dx} = \sum_{n=1}^{\infty} a_n n x^{n-1},$$

$$2x\frac{dy}{dx} = 2\sum_{n=1}^{\infty} n a_n x^n,$$

$$\frac{d^2y}{dx^2} = \sum_{n=2}^{\infty} a_n n(n-1) x^{n-2},$$

$$(1-x^2)\frac{d^2y}{dx^2} = \sum_{n=2}^{\infty} a_n n(n-1) x^{n-2} - \sum_{n=2}^{\infty} a_n n(n-1) x^n. \tag{6.4}$$

Inserting in (6.3) and equating powers of x we must have that

$$(n+2)(n+1)a_{n+2} = -(l^2+l-n^2-n)a_n,$$
$$= -(l-n)(l+n+1)a_n,$$
$$\Rightarrow a_{n+2} = \frac{n(n+1)-l(l+1)}{(n+2)(n+1)}a_n. \tag{6.5}$$

The general solution to (6.3) is the sum of two series containing two constants a_0 and a_1

$$y_1(x) = a_0\left[1 - l(l+1)\frac{x^2}{2} + (l-2)l(l+1)(l+3)\frac{x^4}{4} + \cdots\right],$$

$$y_2(x) = a_1 \left[x - (l-1)(l+2)\frac{x^3}{3} + (l-3)(l-1)(l+2)(l+4)\frac{x^5}{5} + \cdots \right].$$
(6.6)

Since y_1 only contains even powers and y_2 only odd powers they cannot be proportional to each other and must be linearly independent. Thus, the general solution for $|x| < 1$ must be

$$y(x) = b_1 y_1(x) + b_2 y_2(x).$$
(6.7)

In many applications l is an integer. In that case

$$a_{l+2} = \frac{[l(l+1) - l(l+1)]}{(l+1)(l+2)} = 0$$

and we obtain a polynomial of order l. In particular, if l is even $y_1(x)$ is a polynomial and if l is odd $y_2(x)$ is a polynomial. So we may write the general solution (6.7)

$$y(x) = b_l P_l(x) + c_l Q_l(x),$$
(6.8)

where $P_l(x)$ is the polynomial and Q_l corresponds to the other linearly independent solution. If we now demand that $P_l(1) = 1$ we have a set of polynomials (the Legendre polynomials),

$$
\begin{aligned}
P_0(x) &= 1, \\
P_1(x) &= x, \\
P_2(x) &= \frac{1}{2}(3x^2 - 1), \\
P_3(x) &= \frac{1}{2}(5x^3 - 3x), \\
P_4(x) &= \frac{1}{8}(35x^4 - 30x^2 + 3), \\
P_5(x) &= \frac{1}{8}(63x^5 - 70x^3 + 15x), \\
&\vdots
\end{aligned}
$$
(6.9)

The $Q_l(x)$ solutions are functions not polynomials.

Theorem 6.9 $\int_{-1}^{1} P_m(x) P_n(x) = 0$ *if* $m \neq n$.

Proof.

$$
\begin{aligned}
P_m(x)\frac{d}{dx}\left[(1-x^2)\frac{dP_n(x)}{dx}\right] + n(n+1)P_n(x)P_m(x) &= 0, \\
P_n(x)\frac{d}{dx}\left[(1-x^2)\frac{dP_m(x)}{dx}x\right] + m(m+1)P_n(x)P_m(x) &= 0,
\end{aligned}
$$

$$\Rightarrow \frac{d}{dx}\left((1-x^2)[P_m(x)\frac{dP_n(x)}{dx} - P_n(x)\frac{dP_m(x)}{dx}]\right) +$$

$$[n(n+1) - m(m+1)]P_n(x)P_m(x) = 0. \qquad (6.10)$$

Now integrate from -1 to 1, the first term will be zero since $(1-x^2) = 0$ at both limits and since $m \neq n$ we have the result. $\qquad \square$

Generating Function

Definition 6.10

$$\Phi(x,h) = \frac{1}{\sqrt{1 - 2hx + h^2}}, \quad |h| < 1$$

is called the generating function of the Legendre polynomials.

Theorem 6.11

$$\Phi(x,h) = \sum_{l=0}^{\infty} h^l P_l(x).$$

$$
\begin{aligned}
\text{Let } y &= 2xh - h^2, \\
\Phi(x,h) &= (1-y)^{-\frac{1}{2}}, \\
&= 1 + \frac{y}{2} + \frac{\frac{3}{4}}{2}y^2 + \cdots \\
&= 1 + \frac{1}{2}(2xh - h^2) + \frac{3}{8}(2xh - h^2)^2 + \cdots \\
&= 1 + xh - \frac{1}{2}h^2 + \frac{3}{8}(4x^2h^2 - 4xh^3 + h^4) + \cdots \\
&= 1 + xh + h^2(\frac{3}{2}x^2 - \frac{1}{2}) + \cdots \\
&= P_0(x) + hP_1(x) + h^2 P_2(x) + \cdots
\end{aligned}
$$

For a more complete formal proof, see [13]. The generating function allows us to derive some important relations:

$$\frac{\partial \Phi(x,h)}{\partial h} = -\frac{1}{2}(1 - 2xh + h^2)^{-\frac{3}{2}}(-2x + 2h),$$

$$\Rightarrow (1 - 2xh + h^2)\frac{\partial \Phi(x,h)}{\partial h} = (x - h)\Phi,$$

$$\Rightarrow (1 - 2xh + h^2)\sum_{l=1}^{\infty} lh^{l-1}P_l(x) = (x-h)\sum_{l=0}^{\infty} h^l P_l(x).$$

Equating powers of h we find the recurrence relation:

$$lP_l(x) - 2x(l-1)P_{l-1}(x) \ + \ (l-2)P_{l-2}(x) = xP_{l-1}(x) - P_{l-2}(x),$$
$$lP_l(x) \ = \ (2l-1)xP_{l-1}(x) - (l-1)P_{l-2}(x). \qquad (6.11)$$

For each n the second solution $Q_n(x)$ satisfies the same differential equation and also the same recurrence relation:

$$lQ_l(x) \ = \ (2l-1)xQ_{l-1}(x) - (l-1)Q_{l-2}(x),$$
$$(l+1)Q_{l+1} \ = \ (2l+1)xQ_l(x) - lQ_{l-1}(x),$$
$$(2l+1)xQ_l(x) \ = \ lQ_{l-1}(x) + (l+1)Q_{l+1}(x). \qquad (6.12)$$

However, it is singular at $x = \pm 1$. For $|x| \neq 1$ and it can be shown that [13]

$$Q_0(x) \ = \ \frac{1}{2}\ln\left(\frac{1+x}{1-x}\right),$$
$$Q_1(x) \ = \ \frac{x}{2}\ln\left(\frac{1+x}{1-x}\right) - 1,$$
$$Q_2(x) \ = \ \frac{3x^2-1}{4}\ln\left(\frac{1+x}{1-x}\right) - \frac{3x}{2}$$
$$\vdots \qquad\qquad\qquad\qquad (6.13)$$

The Numerical Generation of the Legendre Functions

For the Legendre functions of the first kind (i.e., the Legendre polynomials) then we can simply use the recurrence relation (6.11) starting with $P_0(x) = 1$, $P_1(x) = x$ and work up to larger l values. This procedure is generally stable.

Unfortunately, for the Legendre function of the second kind using the forward recurrence relation is not numerically stable and is subject to cancellation errors. As we have seen, the Q_l satisfy the recurrence relation

$$(l+1)Q_{l+1}(x) \ = \ (2l+1)xQ_l(x) - lQ_{l-1}(x),$$
$$Q_0(x) \ = \ \frac{1}{2}\ln\left[\frac{x+1}{x-1}\right],$$
$$Q_1(x) \ = \ \frac{x}{2}\ln\left[\frac{x+1}{x-1}\right] - 1.$$
$$|x| \ \neq \ 1. \qquad\qquad (6.14)$$

Noting that [20, 21]

$$|Q_l(x)| \ \leq \ e^{-l-1}Q_0(\cosh 2\alpha), \qquad (6.15)$$

where $x = \cosh\alpha, \alpha \geq 0$. Then you can use (6.15) to estimate the value, l_0 say, for which

$$|Q_{l_0}| < 10^{-7}. \qquad (6.16)$$

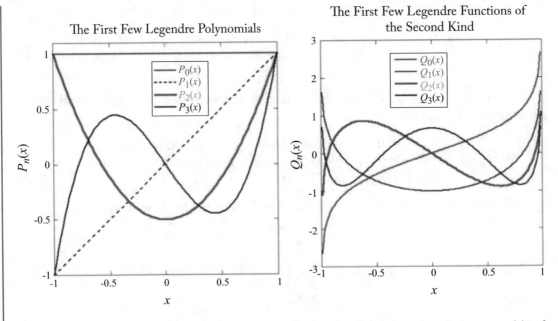

Figure 6.3: The first few Legendre functions of the first kind, left panel, and the second kind, right panel.

Then set

$$\begin{aligned}
\tilde{Q}_{l_0}(x) &= 0, \\
\tilde{Q}_{l_0-1}(x) &= 10^{-7}.
\end{aligned} \tag{6.17}$$

Then calculate $\tilde{Q}_l$ using the bacward recurrence formula for $0 \leq l \leq l_0$. Normalize the sequence $\tilde{Q}_l(x)$ to deduce the computed Legendre functions $Q_l(x)$ using the analytic $Q_0(x)$

$$Q_l(x) = \frac{Q_0(x)}{\tilde{Q}_0(x)} \tilde{Q}_l(x).$$

The error in this case is approximately:

$$\left(\frac{Q_0(x)}{\tilde{Q}_0(x)} \right) \times 10^{-7}.$$

Clearly, the powers of ten are arbitrary and could be chosen differently but due care must be taken to work within the precision of the machine. The Legendre polynomials are a member of class of polynomial approximations which are widely used. Their strengths, weaknesses, and utility are most clearly seen when we cast our analysis in terms of vector space theory.

6.3   INFINITE DIMENSIONAL VECTOR SPACES

Not all vector spaces one uses are finite dimensional. In particular, the vector space of states from Quantum Mechanics is infinite dimensional.[1] As an example, consider the set of piecewise continuous functions, $\mathbb{PC}[a,b]$ on the real interval $[a,b]$. If $f, g \in \mathbb{PC}[a,b]$ and α, β numbers then $\alpha f + \beta g \in \mathbb{PC}[a,b]$. It is immediately obvious that $\mathbb{PC}[a,b]$ is a vector space over the real numbers. To make $\mathbb{PC}[a,b]$ into a normed linear space we will need an inner product. Let us try the following.

Definition 6.12

$$
\begin{aligned}
f, g &\in \mathbb{PC}[a,b|w] \\
\langle f|g \rangle &= \int_a^b f(x)g(x)w(x)dx, \\
w(x) &> 0,
\end{aligned}
$$

where the "*weight function*," w, is continuous and always positive on $[a,b]$.

Now, clearly, with this definition for $f, g, h \in \mathbb{PC}[a,b|w], \alpha, \beta, \gamma \in \mathbb{R}$

$$
\begin{aligned}
\langle f|g \rangle &= \langle g|f \rangle, \\
\langle f\alpha f + \beta g|\gamma h \rangle &= \alpha\gamma\langle f|h \rangle + \beta\gamma\langle g|h \rangle.
\end{aligned}
\tag{6.18}
$$

Suppose we define a function f_0 which has the value 17 at 10,000 equally spaced points between a and b but is zero otherwise. Since we can always change the value of an integrand at a finite number of points without changing the value of the integral we must have

$$
\int_a^b f_0^2(t)w(t)dt = 0
$$

but the function $f_0 \neq 0$ at 10,000 points in $[a,b]$. At first sight it looks like we can't define a norm. However, we can rescue the situation by agreeing to a distinction between the "*vector*" f_0 and the function f_0.

Definition 6.13   We shall take two elements of the vector space $\mathbb{PC}[a,b|w]$, f, g to be equal if

$$
\langle f - g|f - g \rangle = 0
$$

even if $f(t) \neq g(t)$ at a finite number of points.

With this agreement we have a normed linear space and can define a norm

$$
\|f\|^2 = \langle f|f \rangle = \int_a^b f^2(t)w(t)dt,
$$

---

[1]Technically it is a Hilbert space, $\mathfrak{H}$, i.e., an infinite dimensional vector space with an inner product and associated norm, with the additional property that if $\{x_N\}$ is a sequence in $\mathfrak{H}$ and the difference $\|x_n - x_m\|$ can be made arbitrarily small for n, m big enough then x_N must converge to a limit contained in $\mathfrak{H}$.

with this definition we can consider limits.

Definition 6.14

$$\lim_{n \to \infty} f_n \to f$$

if given $\epsilon > 0$ no matter how small there exists an N s.t. for all $n > N$

$$\|f - f_n\| < \epsilon.$$

Let us consider the set $\mathbb{PC}[-1, 1]$ and define our weight function as

$$w(t) \equiv 1.$$

Now, we know that if we have a power series s.t.

$$\sum_{j=0}^{n} c_j x^j = 0, \tag{6.19}$$

then $c_j = 0$ for all j . Thus, the set is $\mathbb{X} = \{x^j\}_{j=0}^n$ is a collection of linearly independent vectors. Now, I will apply the Grahm–Schmidt orthogonalization process (Lemma A.9) to $\mathbb{X}$ with weight $w \equiv 1$ to create an new set of orthonormal polynomials $\{\phi_n(x)\}$. I will adopt the notation of (A.6). Then,

$$
\begin{aligned}
e_0' &= 1 \\
\langle e_0' | e_0' \rangle &= 2 \\
e_0 &= \phi_0(x) = \sqrt{\frac{1}{2}} \\
e_1' &= x - \frac{1}{2} \int_{-1}^{1} x\,dx \\
&= x, \\
e_1 &= \phi_1(x) = \sqrt{\frac{3}{2}}x, \\
e_2 &= \phi_2(x) = \sqrt{\frac{5}{2}} \left(\frac{3}{2}x^3 - \frac{1}{2} \right), \\
e_3 &= \phi_3(x) = \sqrt{\frac{7}{2}} \left(\frac{5}{2}x^3 - \frac{3}{2}x \right), \\
&\quad \vdots
\end{aligned}
\tag{6.20}
$$

In fact, the set of polynomials I have constructed are directly related to the Legendre polynomials

$$\phi_n(x) = \sqrt{\frac{2n + 1}{2}} P_n(x). \tag{6.21}$$

Table 6.1: Different polynomials bases

Name	Interval	Weight
Legendre	(-1, 1)	1
Laguerre	(0, ∞)	$x^\alpha e^{-x}$
Hermite	(-∞, ∞)	e^{-x^2}
Chebyshev	(-1, 1)	$(1 - x^2)^{-1/2}$
Jacobi	(-1, 1)	$(1 - x)^\alpha (1 + x)^\beta$

The Legendre polynomials are orthogonal and differ only in norm from the set we got by using the Grahm–Scmidt process. They satisfy

$$\int_{-1}^{1} P_n(x) P_m(x) dx = \frac{2}{2n+1} \delta_{nm}, \tag{6.22}$$

and if f is any function in $\mathbb{PC}[-1, 1|1]$ then f may be expanded

$$f(x) = \sum_{n=0}^{\infty} a_n P_n(x), \text{ where}$$

$$a_n = \frac{2n+1}{2} \int_{1}^{1} f(x) P_n(x) dx. \tag{6.23}$$

Remember convergence in the norm might not be equivalent to point-wise convergence. There are other sets of orthogonal polynomials as well as the Legendre, defined on different ranges and with different weight functions; I list some of the most important ones in Table 6.1; for more details of their properties see [13].

6.3.1   ZEROS OF ORTHOGONAL POLYNOMIALS

Let $\{p_m(x)\}$ be a set of polynomials orthogonal over (a, b) with respect to the weight $w(x) > 0$, i.e.,

$$\int_{a}^{b} p_m(x) p_n(x) w(x) dx = 0, \tag{6.24}$$

unless $m = n$.

Lemma 6.15   *The mth polynomial has exactly m zeros, which are simple and lie in (a, b).*

Proof. Suppose $p_m(x)$ changes sign n times in (a, b). Now from the fundamental theorem of algebra $n \leq m$. Let $a_1, \ldots a_n$ be the distinct points where p_m changes sign.

$$q_n(x) = \prod_{i=1}^{n}(x - a_i)$$

is a polynomial of order n and

$$\int_a^b q_n(x)p_m(x)dx = 0, \quad n < m. \tag{6.25}$$

But in the vicinity of a_1 $(x - a_1)p_m(x)$ does not change sign, indeed $(x - a_1).(x - a_2)\cdots(x - a_n)p_m(x)$ does not change sign in (a, b), hence

$$\int_a^b (x - a_1).(x - a_2)\cdots(x - a_n)p_m(x)w(x)dx \neq 0. \tag{6.26}$$

This is in contradiction to (6.25) if n is not equal to m. So there are m distinct zeros. The point here is we have shown there are m sign changes and m is the maximum number of roots a polynomial of order m can have! $\square$

6.4   QUADRATURE

6.4.1   SIMPSON REVISITED

The Simpson rule formula (2.28)

$$\int_{-1}^{1} f(x)dx = \frac{1}{3}[f(-1) + 4f(0) + f(1)]$$

can be written

$$\begin{aligned}
\int_{-1}^{1} f(x)dx &= \frac{1}{3}[f(-1) + 4f(0) + f(1)], \\
&= \sum_{i=1}^{3} \bar{w}_n f(x_n), \\
\bar{w}_1 &= \frac{1}{3} = \bar{w}_3, \\
\bar{w}_2 &= \frac{4}{3}, \\
x_1 &= -1, \\
x_0 &= 0, \\
x_3 &= 1, \tag{6.27}
\end{aligned}$$

and is exact for all polynomials of order less than or equal 3. Be careful not to confuse the weight function w used in the definition of the inner product (Definition 6.12) and the numbers $\bar{w}_n$ which are also called weights in (6.27).

6.4.2   WEIGHTS AND NODES

We want to integrate

$$\int_a^b f(x)w(x)dx.$$

We want to find a set of points $\{x_i\}_{i=0}^n$ and weights $\bar{w}_n$ such that

$$\int_a^b f(x)w(x)dx \approx \sum_0^n \bar{w}_i f(x_i) \tag{6.28}$$

is as near to exact as possible.

Let us begin by assuming we have points $\{x_i\}_{i=0}^3$ and then chose our weights in such away that the integral over linear quadratic and cubic polynomials is exact

$$
\begin{aligned}
f(x) &= 1, \\
\int_{-1}^1 dx &= 2 = \bar{w}_0 + \bar{w}_1 + \bar{w}_2 + \bar{w}_3, \\
f(x) &= x, \\
\int_{-1}^1 x &= 0 = \bar{w}_0 x_0 + \bar{w}_1 x_1 + \bar{w}_2 x_2 + \bar{w}_3 x_3, \\
f(x) &= x^2, \\
\int_{-1}^1 x^2 &= \frac{2}{3} = \bar{w}_0 x_0^2 + \bar{w}_1 x_1^2 + \bar{w}_2 x_2^2 + \bar{w}_3 x_0^2, \\
f(x) &= x^3, \\
\int_{-1}^1 x^3 &= 0 = \bar{w}_0 x_0^3 + \bar{w}_1 x_1^3 + \bar{w}_2 x_2^3 + \bar{w}_3 x_0^3.
\end{aligned}
$$

$$
\begin{bmatrix}
1 & 1 & 1 & 1 \\
x_0 & x_1 & x_2 & x_3 \\
x_0^2 & x_1^2 & x_2^2 & x_3^2 \\
x_0^3 & x_1^3 & x_2^3 & x_3^3
\end{bmatrix}
\begin{pmatrix}
\bar{w}_0 \\
\bar{w}_1 \\
\bar{w}_2 \\
\bar{w}_3
\end{pmatrix}
=
\begin{pmatrix}
2 \\
0 \\
\frac{2}{3} \\
0
\end{pmatrix},
$$

which we can summarize in matrix form as

$$\boldsymbol{Vw} = \boldsymbol{y}.$$

So once we have chosen our $x_i's$ e can chose our $\bar{w}_i's$ by inverting $\boldsymbol{V}$.

$\boldsymbol{V}$ is essentially the Vandemonde matrix that we met doing interpolation. In fact this is the transpose of the matrix $\boldsymbol{V}$ in (6.2), but this should not unduly concern you since

$$\left(\boldsymbol{V}^T\right)^{-1} = \left(\boldsymbol{V}^{-1}\right)^T.$$

In theory, we can just keep looking for a bigger and bigger Vandemonde matrix, We need n weights and nodes to exactly integrate a polynomial of order n but if we make it "*too big*" for an arbitrary set of x_i the weights vary wildly and the sum in (6.28) becomes numerically unstable.

6.4.3   GAUSSIAN QUADRATURE

Example 6.16   Consider the polynomial:

$$f(x) = 3x^5 + 3x^2 + 3x + 3.$$

Divide $f(x)$ by $\frac{5x^3-3}{2}$:

$$
\begin{array}{r}
\frac{6}{5}x^2 \hspace{3.2cm} \\
5x^3 - 3 \overline{)\ 6x^5 \ + 6x^2 + 6x + 6} \\
-6x^5 + \frac{18}{5}x^2 \hspace{1.6cm} \\
\hline
\frac{48}{5}x^2 + 6x + 6
\end{array}
$$

$$p_3(x) = \frac{5x^3 - 3}{2}.$$

So

$$f(x) = p_3(x)q(x) + r(x),$$

where $q(x), r(x)$ are polynomials of order 2.

In the same way if $f(x)$ is a polynomial of order no more than $2n - 1$ we can write

$$f(x) = p_n(x)q(x) + r(x),$$

where q and r are polynomials of order $n - 1$ or less. Suppose now that p_n the polynomial of order n is a member of set of polynomials orthogonal over (a, b) with respect to the weight $w(x) > 0$. We want to find the weights and nodes such that

$$\int_a^b f(x)w(x)dx = \sum_{i=0}^n \bar{w}_i f(x_i).$$

Now, if $m < n$ p_n is orthogonal to the subspace of mononomials constructed from $x^0, x, \ldots, x^m$, hence

$$\int_{-a}^b p_n(x)q(x)w(x)dx = 0. \tag{6.29}$$

If we chose x_i to be the n zeros of the p_n function then

$$\sum_{i=0}^n \bar{w}_i p_n(x_i)q(x_i)$$

is exactly zero the same as (6.29). $r(x)$ is a polynomial of order $n - 1$ so we can find the n weights $\bar{w}_i$ so that

$$\int_1^1 r(x)w(x)dx = \sum_{i=0}^{n-1} \bar{w}_i r(x_i)$$

is exact. Therefore,

$$\int_a^b f(x)w(x)dx = \int_a^b w(x)\left[p_n(x)q(x)+r(x)\right]dx = \sum_{i=0}^{n-1}\bar{w}_i r(x_i) = \sum_{i=0}^{n}\bar{w}_i f(x_i).$$

We know the nodes x_i and now we will look for a closed form expression for the weights. Now since $r(x)$ is a polynomial of order less than n, it is thus fixed by the values it attains at n different points and we can use our Lagrange interpolation to write

$$r(x) = \sum_{i=1}^{n} L_i(x)r(x_i),$$

$$\Rightarrow \int_a^b w(x)r(x)dx = \sum_{i=0}^{n-1}\left[\int_a^b L_i(x)w(x)dx\right]r(x_i),$$

$$\Rightarrow \bar{w}_i = \int_a^b w(x)L_i(x)dx. \tag{6.30}$$

Note that the weights $\bar{w}_i$ and nodes x_i depend only on the polynomial p_n and not on the function f we want to integrate and thus can be tabulated. The choice the polynomial basis we might want to use depends on the interval (a,b) and the integral. Suppose, for example, you wanted to evaluate an integral of the form

$$\int_{-1}^{1}\frac{f(x)}{\sqrt{1-x^2}}dx,$$

then the optimum choice is a Gauss–Chebyshev integration, see Table 6.1, where the weight function is

$$w(x) = \frac{1}{\sqrt{1-x^2}}.$$

CHAPTER 7

Sturm–Liouville Theory

In classical as well as quantum physics many problems arise in the form of boundary value problems involving second order ordinary differential equations, very frequently these problems are of Sturm–Liouville type. Such problems have a particular place in the quantum theory because of the self-adjoint nature of the differential operators involved.

Definition 7.1   A second order differential equation of the form:

$$\frac{d}{dx}\left[p(x)\frac{dy(x)}{dx}\right] + q(x)y(x) \quad = \quad -\lambda w(x)y(x),$$

$$x \quad \epsilon \quad [a,b],$$

with p, q, and w specified and $p(x) > 0, w(x) > 0$ on $[a,b]$ is said to be a Sturm–Liouville equation.

7.1   EIGENVALUES

Note that both $y(x)$ and λ are unspecified in Definition 7.1 so the solution of the Sturm–Liouville equation is essentially an eigenvalue problem. The Sturm–Liouville differential equation as written down above is a purely formal entity in the absence of boundary conditions. We can define a new Hilbert space of square integral functions on $[a,b]$, $\mathbb{L}^2[a,b|w]$, with an inner product

$$\langle f|g\rangle = \int_a^b \bar{f}(x)g(x)w(x)dx, \tag{7.1}$$

where the "*weight function*" $w(x)$ is real and positive.

Definition 7.2   We define a Sturm–Liouville differential operator, $\hat{L}$ to be an operator

$$\frac{-1}{w(x)}\frac{d}{dx}\left[p(x)\frac{d}{dx}\right] + q(x).$$

In order to properly define $\hat{L}$ on our Hilbert space we need to add boundary conditions. Now if we can find such conditions such that $\hat{L}$ is self adjoint then:

(i) its eigenvalues would be real (Lemma A.24),

(ii) its eigenfunctions corresponding to distinct eigenvalues would be orthogonal (Lemma A.25).

Let us look to see what such boundary conditions would look like. Consider

$$\langle f|\hat{L}g\rangle \;=\; \int_a^b \bar{f}(x)\left[-\frac{d}{dx}[p(x)g'(x)]+q(x)g(x)\right]dx$$

integrating the first term on the right by parts we have

$$\langle f|\hat{L}g\rangle$$
$$=\; -\bar{f}(x)p(x)g'(x)\big|_a^b + \int_a^b \{\bar{f}'(x)p(x)g'(x)+\bar{f}(x)q(x)g(x)\}dx.$$

In the same way,

$$\langle \hat{L}f|g\rangle \;=\; -\bar{f}'(x)p(x)g(x)\big|_a^b + \int_a^b \{\bar{f}'(x)p(x)g'(x)+\bar{f}(x)q(x)g(x)\}dx,$$

$$\Rightarrow \langle f|\hat{L}g\rangle - \langle \hat{L}f|g\rangle \;=\; -\bar{f}(x)p(x)g'(x)\big|_a^b + \bar{f}'(x)p(x)g(x)\big|_a^b.$$

To get a self adjoint operator we would need:

$$fpg'\big|_a^b \;\equiv\; -\bar{f}(b)p(b)g'(b)+\bar{f}'(b)p(b)g(b)+\bar{f}(a)p(a)g'(a)-\bar{f}'(a)p(a)g(a)$$
$$=\; \left[p(x)\left(\frac{d\bar{f}(x)}{dx}g(x)-\bar{f}(x)\frac{dg(x)}{dx}\right)\right]_a^b$$
$$=\; 0. \tag{7.2}$$

If we define our differential operator to be the formal differential operator $\hat{L}$ together with boundary conditions of the form (7.2) then we have a self adjoint operator on $\mathbb{L}^2[a,b|w]$. The boundary conditions (7.2) would be satisfied if, for example, we were to consider functions, g, defined on $[a,b]$ satisfying

$$\begin{aligned}\alpha_1 g'(a) + \alpha_2 g(a) &= 0,\\ \beta_1 g'(b) + \beta_2 g(b) &= 0,\end{aligned} \tag{7.3}$$

α_i and β_i are constants not both zero. It is important to recognize that we must choose the same constants for all our functions. We will describe (7.3) as "*regular*" boundary conditions. If the function $p(x)$ is such that $p(a) = p(b)$ then we can impose alternative "*periodic boundary conditions*"

$$g(a) \;=\; g(b),$$

$$g'(a) \; = \; g'(b).$$

In either case the Sturm–Liouville differential operator $\hat{L}$ is self adjoint with real eigenvalues and orthogonal eigenvectors provided only that the eigenvalues are non degenerate.

But there is more. It can be shown [22] that in the regular case.

- The eigenvalues are simple; in other words the eigenfunctions are non-degenerate and thus mutually orthogonal.

- The set of eigenvalues is countably infinite and can be arranged in a monotonically increasing sequence bounded below:

$$\lambda_0 \; < \; \lambda_1 < \cdots < \lambda_n < \cdots$$
$$\lim_{n \to \infty} \lambda_n \; = \; \infty.$$

- The eigenfunction corresponding to the nth eigenvalue has n zeros on the open interval (a, b).

- The orthonormal set of eigenfunctions form a basis for the Hilbert space.

For the periodic problem and most but not all of the above results will still hold. The eigenvalues will still be real, the eigenfunctions orthogonal for different eigenvalues but with the exception that these may be degeneracy.

Consider the periodic Sturm–Liouville problem

$$\frac{-\hbar^2}{2m} \frac{d^2 \psi(x)}{dx^2} \; = \; E\psi,$$
$$\psi(0) \; = \; \psi(L),$$
$$\psi'(0) \; = \; \psi'(L).$$

This has eigenvalues

$$E_n = \frac{4\pi^2 \hbar^2 n^2}{2mL^2},$$

and for each $E_n, n \neq 0$ there are two linearly independent solutions

$$\cos(\frac{2\pi n x}{L}), \quad \sin(\frac{2\pi n x}{L}).$$

We can always use our Grahm–Schmidt procedure to find orthogonal vectors that span the subspace corresponding to a degenerate eigenvalue, so even in the periodic case we can find an othonormal set of eigenfunctions that form a basis but the sequence of eigenvalues is not monotonically increasing.

7.2 LEAST SQUARES APPROXIMATION

Suppose g is a member of $\mathbb{L}^2[a, b|w]$ then we can expand it in terms of the orthonormal eigen-functions of our Sturm–Liouville operator, $\{\phi_i\}_{m=1}^{\infty}$

$$g(x) = \sum_{n=1}^{\infty} c_n \phi_n(x),$$

$$||g||^2 = \langle g|g \rangle = \sum_{n=1}^{\infty} |c_n|^2. \tag{7.4}$$

Obviously, computers cannot deal with infinite sums. Suppose we approximate g by a function f which contains only a finite number of orthonormal eigenfunctions:

$$f(x) = \sum_{n=1}^{N} b_n \phi_n(x). \tag{7.5}$$

We want to chose the constants b_i s.t. f is the "*best*" approximation to g, just as we did in Chapter 5 we will look for the least squared fit, that is we want to minimize $||f - g||^2$

$$||f - g||^2 = \langle f|f \rangle + \langle g|g \rangle - \langle f|g \rangle - \langle g|f \rangle,$$

$$= \left[\sum_{i=1}^{N} |b_i|^2 - \bar{b}_i c_i - \bar{c}_i b_i \right] + ||g||^2. \tag{7.6}$$

We want to chose our set of N coefficients b_i so that the term in the square brackets is smallest. Write

$$F(\bar{b}, b) = \left[\sum_{i=1}^{N} \bar{b}_i b_i - \bar{b}_i c_i - \bar{c}_i b_i \right].$$

A necessary condition of a minimum is that $F(\bar{b}, b)$ be stationary w.r.t. b_i and $\bar{b}_i$:

$$\frac{\partial F}{\partial b_i} = \bar{b}_i - \bar{c}_i = 0,$$

$$\frac{\partial F}{\partial \bar{b}_i} = b_i - c_i = 0. \tag{7.7}$$

These derivatives vanish if $c_i = b_i$ and further

$$\frac{\partial^2 F}{\partial b_i \partial b_j} = 0,$$

$$\frac{\partial^2 F}{\partial \bar{b}_i \partial \bar{b}_j} = 0,$$

$$\frac{\partial^2 F}{\partial b_j \, \partial \bar{b}_j} \;=\; \delta_{ij} \geq 0. \tag{7.8}$$

The extremum is a minimum! Therefore, if we wish to approximate a function $g(x)$ by representing it as a linear combination of just a finite number of eigenfunctions of some Sturm–Liouville operator, the best we can do is to choose exactly the same coefficients as in the true infinite expansion.

CHAPTER 8

Case Study: The Quantum Oscillator

The harmonic oscillator and the hydrogen atom are two of the very few quantum systems which admit simple analytic solutions. All numerical methods need to be benchmarked against these two cases. In this chapter I am going to show you how the understanding of Sturm–Liouville problems you acquired in the proceeding chapter can be applied to the computation of the oscillator eigenvalues and eigenfunctions.

8.1 NUMERICAL SOLUTION OF THE ONE DIMENSIONAL SCHRÖDINGER EQUATION

Let us consider the Schrödinger equation in one dimension [23]

$$\left[-\frac{\hbar^2}{2m}\frac{d^2\psi(x)}{dx^2} + V(x) \right]\psi(x) = E\psi(x). \tag{8.1}$$

We are looking for "*bound states*," i.e., states where

$$\lim_{x\to\pm\infty}\psi(x) = 0.$$

We want to find both the eigenvalues and the eigenfunctions. As you know an eigenvalue problem can only have solutions for certain values of E. Suppose that E is such a value then the Schrödinger equation (8.1) can be written:

$$\frac{d^2\psi(x)}{dx^2} + k^2(x)\psi(x) = 0,$$
$$k(x) = \sqrt{\frac{2m}{\hbar^2}[E - V(x)]},$$
$$\psi(x) \rightarrow 0 \text{ as } x \to \pm\infty. \tag{8.2}$$

At this early stage it is helpful to focus on (8.2) and to see how much of the character of the solution we can deduce before we start calculating.

We immediately recognize that we have a regular Sturm–Liouville problem, so we expect that the eigenvalues will be real, non-degenerate, bounded below and that they can be ordered

$$E_0 < E_1 < E_2 < \cdots \tag{8.3}$$

Further, the eigenfunction corresponding to the nth eigenvalue will have exactly n zeros. If the potential is symmetric $V(x) = V(-x)$ then we make use of the following result.

Lemma 8.1   *Suppose the potential V is such that*

$$V(-x) = V(x)$$

and each bound state level corresponds to only one independent solution then

$$\psi(-x) = \pm\psi(x).$$

If we have the positive sign then we say the wave function has even parity and if negative we say that we have odd parity.

Proof. Suppose $\psi(x)$ is a solution of the Schrödinger equation corresponding to the energy E then

$$-\frac{\hbar^2}{2m}\frac{d^2\psi(x)}{dx^2} + V(x)\psi(x) = E\psi(x). \tag{8.4}$$

Now we can always replace x by $-x$ in (8.4) and remembering that $V(-x) = V(x)$ we see that $\psi(-x)$ is also a solution corresponding to the same eigenvalue and since the eigenvalues are non degenerate it follows that $\psi(x)$ and $\psi(-x)$ must be linearly dependent, i.e.,

$$\psi(x) = C\psi(-x),$$

for all x. Replacing x by $-x$ once more yields

$$\psi(x) = C^2\psi(x),$$

hence $C^2 = 1$, therefore

$$\psi(-x) = \pm\psi(x). \tag{8.5}$$

$\square$

For a given value of the energy E, we can divide space into three regions, depending on the value of $k(x)$ in (8.2). This has something of the character of the classical problem we discussed in Chapter 4.

In region 2,

$$E \geq V,$$
$$k^2(x) > 0,$$

and we would expect to find an oscillatory solution but in regions 1 and 3

$$E < V,$$

$$k^2 \; < \; 0.$$

Asymptotically, we expect to find an exponentially decaying solution, see Appendix B. We expect to find the nodes of the wave function concentrated in region 2. Classically, there will be no solution for regions 1 and 3. The points, x_{tp} for which

$$E - V(x_{tp}) = 0 \qquad (8.6)$$

marks the "*turning points*" between the classically allowed and quantum regions. We can make use of our knowledge of Sturm–Liouville equations to create a computer code to get an estimate of the eigenvalues and eigenfunctions. We know that the eigenfunction corresponding to the nth eigenvalue has n zeros, smaller eigenvalues will have less zeros, bigger eigenvalues will have more. We expect to find all the zeros in region 2. Our potential is symmetric, therefore we know that the eigenfunctions will be either symmetric or antisymmetric either way if $x_0 > 0$ is a nodal point then $-x_0$ is also a nodal point. Further, if $\psi(x)$ has odd parity:

$$
\begin{aligned}
\psi(-h) &= -\psi(h), \\
\psi(-h) + \psi(h) &= 0, \\
\Rightarrow 2\psi(0) + O(h^2) &= 0, \\
\Rightarrow \psi(0) &= 0, \qquad (8.7)
\end{aligned}
$$

if $\psi(x)$ has even parity

$$
\begin{aligned}
\psi(-h) &= \psi(h), \\
\frac{\psi(h) - \psi(-h)}{h} &= 0, \\
\Rightarrow \psi'(0) &= 0. \qquad (8.8)
\end{aligned}
$$

For either m even or odd, we will have exactly the same number of zeros for $x > 0$ and $x < 0$. If ψ is an odd function it must, as we have just seen, have a zero at the origin so it must have an odd number of zeros and if ψ is an even function it must have an even number of zeros. (If it had an odd number of nodes, there must be one node at the origin, since there is exactly the same number of nodes for $x > 0$ and $x < 0$, so $\psi(0) = 0 = \psi'(0)$ therefore the leading term in the Taylor's expansion is just: $x^2 \psi''(0) = k^2(0)\psi(0) = 0$ and so on from the higher terms, i.e., function is exactly zero.)

In summary:

- if the function has m nodes and is odd then there must be one node at the origin and $\frac{m-1}{2}$ nodes for $x > 0$,

- if the function has m nodes and is even then there must be $\frac{m}{2}$ nodes for $x > 0$,

either way we need only solve for $x \geq 0$.

8.2   NUMERICAL SOLUTION FOR THE OSCILLATOR

While the quantum oscillator problem admits a relatively simple analytic solution, see Appendix B, our ambition here is to find an efficient numerical approach to calculate the eigenfunctions and eigenvalues.

The simplest approach is called the "*shooting method.*" It searches for a function with with a pre-determined number, n, of zeros. It is assumed that the actual eigenvalue E_n lies somewhere in an energy range $[E_{min}, E_{max}]$. An energy E is taken to be

$$E = \frac{E_{max} + E_{min}}{2}. \tag{8.9}$$

The energy range should contain the desired eigenvalue E_n. The wave function is integrated starting from $x = 0$ in the direction of positive x; at the same time, the number of nodes, m (i.e., of changes of sign of the function) is counted. If the number of nodes is larger than n, E is too high; if the number of nodes is smaller than n, E is too low. A new interval is defined

$$\begin{aligned} E_{max} &= E \quad \text{if } m > n, \\ E_{min} &= E \quad \text{if } m < n. \end{aligned} \tag{8.10}$$

A replacement E is found from (8.9) and the procedure repeated until the energy interval is smaller than a pre-determined threshold, we assume that convergence has been reached.

I wrote a code to study the $n = 3$ eigenvalue and eigenfunction of the harmonic oscillator potential where units were chosen such that $\hbar = m = \omega = 1$ and the range of x was taken to be

$$-10 \leq x \leq 10$$

which was divided into 300 equally spaced intervals. The potential was $V(x)$ was calculated at each grid point and stored an array $V(i)$ then the initial values of E_{max} and E_{min} were determined by

$$\begin{aligned} E_{max} &= \max_i |V(i)|, \\ E_{min} &= \min_i |V(i)|. \end{aligned}$$

Our differential equation is of Numerov form (3.27),

$$\psi''(x) = -g(x)\psi(x) + s(x),$$

where we take

$$\begin{aligned} s(x) &= 0, \\ g(x) &= \frac{2m}{\hbar} [E - V(x)], \end{aligned}$$

Table 8.1: Results from Numerov

Iteration	Energy Eigenvalue
1	25.000000000000000
2	12.500000000000000
3	6.2500000000000000
4	3.1250000000000000
5	4.6875000000000000
6	3.9062500000000000
7	3.5156250000000000
8	3.3203125000000000
9	3.4179687500000000
10	3.4667968750000000
⋮	⋮
35	3.4999996962142177
36	3.4999996954866219
37	3.4999996958504198
38	3.4999996960323188
39	3.4999996961232682

consequently I used the formula (3.36)

$$\psi_{n+1} = \frac{2\psi_n\left(1 - \frac{5h^2}{12}g_n\right) - \psi_{n-1}\left(1 + \frac{h^2}{12}g_{n-1}\right)}{1 + g_{n+1}\frac{h^2}{12}} + O(h^6)$$

to perform the numerical integration.

It was necessary only to integrate from 0 in the direction of positive x. Because I was focused on the $n = 3$ eigenvalue I knew the eigenfunction would have odd parity with its first zero at the origin. Consequently, I took ψ_0 to be 0, $\psi_1 = h$, $\psi_{-1} = -h$. The choice of ψ_1 is actually somewhat arbitrary but is consistent with the lowest order Taylor expansion.[1] Any arbitrariness I include will be remove once I normalize the wavefunction. I took the threshold to be fixed at $|E_{\max} - E_{\min}| < 10^{-10}$. The output is shown in Table 8.1.

The correct eigenvalue 3.5 is recovered quite quickly. In Figure 8.1, the associated eigenfunction is compared with the analytic solution. In the classical accessible region the analytic

---

[1]Had I been interested in an even eigenfunction I would have taken ψ_0 to be arbitrary but finite $\psi_{+1} = \psi_{-1}$ so I could deduce ψ_{+1} from the Numerov formula, (3.36).

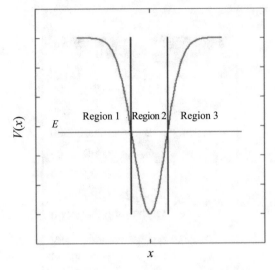

Figure 8.1: Given a test Energy space is divided into 3. The region 1 and 3 with $E \leq V$ are classically forbidden.

and numerical solutions are in moderate agreement but they diverge dramatically once we pass the turning points.

The problem with this approach lies in the fact that the code tried to integrate from $x = 0$ in the region 2 to large x in region 3, but as we have seen the solution to the formal differential equation allows for an exponentially increasing solution which we don't want as well as an exponentially decreasing solution which we do want. If even a tiny amount of the exponentially increasing solution (due to numerical noise, for instance) is present at the turning point, the integration algorithm will inexorably make it grow in the classically forbidden region. In order to deal with this problem we can go "*far*" into the quantum region where we could reasonably assume the wave function is close to zero and integrate backward to the turning point where we "*match*" to the solution integrated from 0 in the classical region.

At the turning point we require that function got by integrating in, from large x_{max} in region 3, $\psi_3(x)$, be such that it matches the solution got from integrating out from 0, $\psi_2(x)$. Matching means that we require that both the functions and their first derivatives are continuous. If we have found the correct eigenvalue then we have our solution.

A second code was written for the harmonic oscillator. Two integrations were performed: a forward recursion, in region 2, starting from $x = 0$, and a backward one, in region 3, starting from x_{max} The matching point was chosen to be the grid point, itp nearest to x_{tp}. Note that since x_{tp} will vary with the choice of E. The outward integration is performed until grid point itp, yielding a function $\psi_{(2)}(x)$ defined in region 2, of course because of symmetry we only need to integrate from 0 to x_{tp}; the number n of changes of sign is counted in the same way as before.

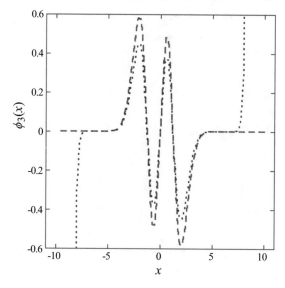

Figure 8.2: Comparison between the analytic solution, red dashed with the numerical solution, blue dotted using the node counting approach.

We note that it is not needed to look for changes of sign beyond x_{tp}: we expect that in the classically forbidden region there will not be any nodes (no oscillations, just exponentially decaying or increasing solutions).

If the number of nodes is the expected one, the code starts to integrate inward from the rightmost points. It goes one grid point beyond x_{max} say, n and then puts

$$\begin{aligned} \psi_{n+1} &= 0 \\ \psi_n &= h \end{aligned}$$

and then use the Numerov formula to integrate to itp. Continuity at this point is easily achieved by simply scaling the solution in region 3 by

$$\frac{\psi_{(2)}(itp)}{\psi_{(3)}(itp)}.$$

Forcing the two solutions to have identical first derivatives is a little more demanding. If we use our Taylor expansion on both functions then

$$\begin{aligned} \psi_{(3)}(itp+1) &= \psi_3(itp) + \psi_3'(itp)h + \frac{1}{2}\psi_3''(itp)h^2 + \mathcal{O}(h^3) \\ \psi_{(2)}(itp-1) &= \psi_{(2)}(itp) - \psi_{(2)}'(itp)h + \frac{1}{2}\psi_{(2)}''(itp)h^2 + \mathcal{O}(h^3). \end{aligned} \qquad (8.11)$$

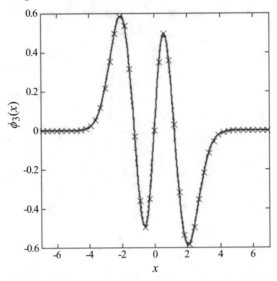

Figure 8.3: Comparison between the analytic solution, red solid with the numerical solution using the extended approach to $\mathcal{O}(h^3)$, blue crosses, for the $n = 3$ eigenfunction of the quantum oscillator.

Now by construction,

$$\psi_{(3)}(itp) \;=\; \psi_{(2)}(itp) = \psi(itp),$$

and from Numerov equation we have

$$\psi''_{(3)}(itp) \;=\; -\psi(itp)g_{itp} = \psi''_{(2)}(itp).$$

Therefore to $\mathcal{O}(h^2)$,

$$\psi'_{(3)}(itp) - \psi'_{(2)}(itp) \;=\; \frac{\psi_{(2)}(itp-1) + \psi_{(3)}(itp+1) - [2 - g_{itp}h^2]\psi(itp)}{h}.$$

$$(8.12)$$

The jump condition (8.12) depends on our choice of E both through the functions $\psi_{(2)}$ and $\psi_{(3)}$ and g_{itp}, and thus (8.12) can be solved for zero difference by successive bisections and this was the approach incorporated in the second code.

In Figure 8.3, a comparison is presented between the analytic solution and the numerical solutions using the modified code, the numerical results of which are visually indistinguishable from the analytic result.

C H A P T E R   9

Variational Principles

Minimization principles have a special place in numerical methods and they form one of the most wide-ranging means of formulating mathematical models governing the equilibrium configurations of physical systems.

9.1   RAYLEIGH–RITZ THEOREM

Theorem 9.1   *Suppose $\hat{H}$ is a self-adjoint operator on a Hilbert space, whose eigenfunctions, ψ_n, form a normalized basis and whose eigenvalues are bounded below and can be ordered:*

$$E_0 < E_1 \leq E_2 \leq \cdots \leq E_n \leq E_{n+1} \leq \cdots,$$

where the smallest eigenvalue is non-degenerate.

Suppose ψ is some arbitrary normalized state then

$$\langle \psi | \hat{H} \psi \rangle \geq E_0,$$

with equality iff $\psi = \psi_0$.

Proof. Since $\hat{H}$ is self adjoint its eigenvalues are real and the basis of eigenfunctions can be chosen such that:

$$\langle \psi_n | \psi_m \rangle = \delta_{nm}.$$

Since ψ_n form a basis and ψ has unit norm

$$
\begin{aligned}
\psi &= \sum_n c_n \psi_n, \\
\langle \psi | \psi \rangle &= \sum_{mn} \bar{c}_m c_n \langle \psi_m | \psi_n \rangle, \\
&= \sum_n |c_n|^2 = 1.
\end{aligned}
$$

Hence,

$$\langle \psi | \hat{H} \psi \rangle = \sum_{n=0,m=0}^{\infty} \bar{c}_m c_n \langle \psi_m | \hat{H} \psi_n \rangle$$

$$
\begin{aligned}
&= \sum_{n=0,m=0}^{\infty} \bar{c}_m c_n E_n \langle \psi_m | \psi_n \rangle \\
&= \sum_{n=0}^{\infty} |c_n|^2 E_n \\
&\geq E_0 \sum_n |c_n|^2 \\
&= E_0, \\
&= E_0 \sum_{n=0}^{\infty} |c_n|^2 + \sum_{n=0}^{\infty} |c_n|^2 [E_n - E_0] \\
&\geq E_0.
\end{aligned}
$$

$\square$

Clearly, the conditions of the theorem applies to any regular Sturm–Liouville problem as we discussed in Chapter 7. We further note that if ψ_0 is the actual normalized state corresponding to E_0 then

$$
\langle \psi_0 | \hat{H} \psi_0 \rangle = E_0, \tag{9.1}
$$

and since $E_n > E_0$ for all $n > 0$ and ψ is any other function then

$$
\langle \psi | \hat{H} \psi \rangle > E_0. \tag{9.2}
$$

This theorem is known as the Rayleigh–Ritz Theorem. The result is not restricted to the finite dimensional case so it is equally valid for any of the infinite dimensional space of functions we have met so far. Further if we know from, experiment say, the value of E_0 then if we can find ψ that minimizes $\langle \psi | \hat{H} \psi \rangle$ we will have found the ground state wavefunction, ψ_0. This observation underlies the various variational approaches to structure studies of many body quantum mechanical systems [23–25].

Now consider a family of states, $\{\psi(\alpha)\}$; depending on real parameters:

$$
\alpha_1, \ldots, \alpha_N,
$$

we can relax our assumption that the states are normalized and define

$$
E(\alpha) = \frac{\langle \psi(\alpha) | H | \psi(\alpha) \rangle}{\langle \psi(\alpha) | \psi(\alpha) \rangle}. \tag{9.3}
$$

$E(\alpha)$ is known as the Rayleigh–Ritz quotient. From Theorem 9.1 we have that $E(\alpha) \geq E_0$ and now we try to make $E(\alpha)$ as small as possible. Now the condition that the real function $E(\alpha)$ be stationary is [11],

$$
\frac{\partial E(\alpha)}{\partial \alpha_i} = 0. \tag{9.4}
$$

Example 9.2   Suppose we want to find the ground state energy of the one-dimensional harmonic oscillator

$$\hat{H} = -\frac{\hbar^2}{2m}\frac{d^2}{dx^2} + \frac{1}{2}m\omega^2 x^2. \tag{9.5}$$

Now we already know the exact answer but lets try out our variational approach. We pick our trial function to be the "*Gaussian*"

$$\psi_T(x) = Ae^{-bx^2}, \tag{9.6}$$

where b is our variational parameter. Our constraint is:

$$
\begin{aligned}
1 &= \int_{-\infty}^{\infty} |A|e^{-2bx^2} dx = 1 \\
&= |A|^2 \sqrt{\frac{\pi}{2b}}, \\
\Rightarrow A &= \left(\frac{2b}{\pi}\right)^{\frac{1}{4}}. \tag{9.7}
\end{aligned}
$$

$$
\begin{aligned}
\langle \psi_T | \hat{H} \psi_T \rangle &= |A|^2[-\frac{\hbar^2}{2m}\int_{-\infty}^{\infty}(e^{-bx^2}\frac{de^{-bx^2}}{dx})dx + \frac{m\omega^2}{2}\int_{-\infty}^{\infty} x^2 e^{-bx^2}dx] \\
&= \frac{\hbar^2 b}{2m} + \frac{m\omega^2}{8b}. \tag{9.8}
\end{aligned}
$$

We have only one free parameter, b

$$
\begin{aligned}
f(b) &= \langle \psi_T | \hat{H}\psi_T \rangle = \frac{\hbar^2 b}{2m} + \frac{m\omega^2}{8b}, \\
\frac{df(b)}{db} &= \frac{\hbar^2}{2m} - \frac{m\omega^2}{8b^2} = 0, \\
\Rightarrow b &= \frac{m\omega}{2\hbar}, \\
\Rightarrow E_T &= \frac{1}{2}\hbar\omega, \\
\Rightarrow \psi_T &= \left(\frac{m\omega}{2\hbar}\right)^{\frac{1}{4}} exp(-\frac{m^2\omega^2}{4\hbar^2}). \tag{9.9}
\end{aligned}
$$

We have unearthed the "*exact*" solution and we can't do any better.

More generally, consider the Schrödinger equation operator in one dimension

$$\hat{H} = -\frac{1}{2}\frac{\hbar^2}{2m}\frac{d^2}{dx^2} + V(x).$$

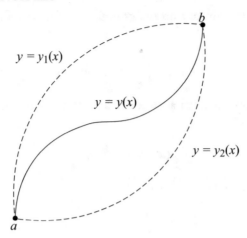

Figure 9.1: The points a and b can be connected by an infinite number of different paths; in this figure we show just 3.

Define

$$
\begin{aligned}
I(\bar{\psi}, \psi) &= \langle \psi | \hat{H} \psi \rangle \\
&= \int_{-\infty}^{\infty} \bar{\psi}(x) \left[-\frac{\hbar^2}{2m} \frac{d^2}{dx^2} + V(x) \right] \psi(x) dx.
\end{aligned}
\tag{9.10}
$$

For any given $\psi(x)$ and $\bar{\psi}(x)$ the integral I in (9.10) is just a number. Our task is to find the ψ out of the infinity of possible ψ's that minimizes $I(\bar{\psi}, \psi)$ while at same time satisfying the constraint that

$$
\langle \psi | \psi \rangle = \int_{-\infty}^{\infty} \bar{\psi}(x) \psi(x) dx = 1.
\tag{9.11}
$$

This is the basic problem from the calculus of variations.

9.2   THE EULER–LAGRANGE EQUATIONS

Suppose F is a function of an independent variable x, a dependent variable $y(x)$, and its derivative $y'(x) \equiv \dfrac{dy(x)}{dx}$. Let $y = f(x)$ define a path, i.e., a curve in two space and suppose the points $a = f(x_a)$ and $b = f(x_b)$ are on this path. Suppose F is a function of y, $\dfrac{dy}{dx}$ and x. The integral

$$
I = \int_{x_a}^{x_b} F(y, \frac{dy}{dx}, x) dx
\tag{9.12}
$$

is just a number for a given path $y(x)$. The question is how to find a particular path $y(x)$ out of the infinity of possible paths which for which I is smallest. Assume $y(x)$ is that path; then define a set of varied curves around it

$$Y(x) = y(x) + \epsilon \eta(x),\tag{9.13}$$

where $\eta(x_a) = \eta(x_b) = 0$ and η is as differentiable as we like.
    Then

$$I(\epsilon) = \int_{x_a}^{x_b} F(Y, \frac{dY}{dx}, x)dx\tag{9.14}$$

defines a function

$$\begin{aligned} I &: \mathbb{R} \to \mathbb{R}, \\ &: \epsilon \to I(\epsilon). \end{aligned}$$

We are assuming that $I(\epsilon)$ takes its extremum value when $\epsilon = 0$, i.e., we want

$$\frac{dI(\epsilon)}{d\epsilon}\bigg|_{\epsilon=0} = 0.\tag{9.15}$$

Now:

$$I(\epsilon) = \int_{x_a}^{x_b} (F(y + \epsilon\eta, y' + \epsilon\eta', x),\tag{9.16}$$

expand in a Taylor series in ϵ

$$I(\epsilon) = \int_{x_a}^{x_b} (F(y, y', x)dx + \int_{x_a}^{x_b} (\frac{\partial F}{\partial y}\epsilon\eta + \frac{\partial F}{\partial y'}\epsilon\eta')dx + O(\epsilon^2).\tag{9.17}$$

Differentiating with respect to ϵ and putting $\epsilon = 0$ we have, for all η,

$$\frac{dI(\epsilon)}{d\epsilon}\bigg|_{\epsilon=0} = \int_{x_a}^{x_b} (\frac{\partial F}{\partial y}\eta + \frac{\partial F}{\partial y'}\eta')dx.\tag{9.18}$$

integrating by parts then

$$\frac{dI(\epsilon)}{d\epsilon}\bigg|_{\epsilon=0} = \int_{x_a}^{x_b} (\frac{\partial F}{\partial y} - \frac{d}{dx}\frac{\partial F}{\partial y'})\eta dx + \eta \frac{\partial F}{\partial y'}\bigg|_{x_a}^{x_b}.\tag{9.19}$$

The second term on the right is zero since $\eta(x_a) = \eta(x_b) = 0$. Since η is arbitrary it follows that a *necessary* condition for an extremum is that

$$\frac{\partial F}{\partial y} - \frac{d}{dx}\frac{\partial F}{\partial y'} = 0.\tag{9.20}$$

(9.20) is known as the Euler–Lagrange equation. There are two special cases when it is particularly simple to solve.

(i) F has no explicit dependence on y i.e.,

$$\frac{\partial F}{\partial y} = 0,$$

$$\Rightarrow \frac{\partial F}{\partial y'} = constant \tag{9.21}$$

(ii) F has no explicit dependence on x i.e., $F = F(y, y')$ then

$$F - y'\frac{\partial F}{\partial y'} = constant. \tag{9.22}$$

The result (9.20) can be extended in a straightforward manner to more than one dependent variable.

Suppose

$$F = F(y_1, ..., y_n, y'_1, ..., y'_n, x).$$

Then the required path $y(x)$ which yields extrema satisfies the set of equations

$$\frac{\partial F}{\partial y_j} - \frac{d}{dx}\left(\frac{\partial F}{\partial y'_j}\right) = 0 \ \ 1 \leq j \leq n. \tag{9.23}$$

Example 9.3   Consider a particle, mass m moving in space under the effect of a scalar potential $V(x, y, z)$ then define the Lagrangian

$$L = \frac{1}{2}m(\dot{x}^2 + \dot{y}^2 + \dot{z}^2) - V(x, y, z). \tag{9.24}$$

Then, requiring the integral

$$I = \int_{t_1}^{t_2} L(x, y, z, \dot{x}, \dot{y}, \dot{z}) \tag{9.25}$$

to be stationary leads to the Euler–Lagrange equations:

$$\frac{d}{dt}\left(\frac{\partial L}{\partial \dot{x}}\right) - \frac{\partial L}{\partial x} = 0 \Rightarrow m\frac{d\dot{x}}{dt} = -\frac{\partial V}{\partial x},$$

$$\frac{d}{dt}\left(\frac{\partial L}{\partial \dot{y}}\right) - \frac{\partial L}{\partial y} = 0 \Rightarrow m\frac{d\dot{y}}{dt} = -\frac{\partial V}{\partial y},$$

$$\frac{d}{dt}\left(\frac{\partial L}{\partial \dot{z}}\right) - \frac{\partial L}{\partial z} = 0 \Rightarrow m\frac{d\dot{z}}{dt} = -\frac{\partial V}{\partial z} \tag{9.26}$$

which is equivalent to the vector equation:

$$\frac{d\boldsymbol{p}}{dt} = -\nabla V. \tag{9.27}$$

which we recognize as Newton's second law.

This is a special case of the *"Principle of Least Action"* [11, 17].

THE PRINCIPLE OF LEAST ACTION

If a classical mechanical system specified by *"generalized coordinates"*

$$q_1(t), \ldots, q_N(t),$$

with kinetic energy, $T(\boldsymbol{q}_i, \dot{\boldsymbol{q}}_i)$ and potential energy $V(\boldsymbol{q}_i, t)$ then the motion of the system from time t_1 to time t_2 is such as to render the *"action integral"*

$$I = \int_{t_1}^{t_2} L(\boldsymbol{q}_i, \dot{\boldsymbol{q}}_i, t) dt$$

stationary, where L is the *"Lagrangian"* defined by

$$L((\boldsymbol{q}_i, \dot{\boldsymbol{q}}_i, t) = T(\boldsymbol{q}_i, \dot{\boldsymbol{q}}_i) - V(\boldsymbol{q}_i, t).$$

A few comments.

- The name is a bit misleading. I didn't have to require the integral have a minimum to recover Newton's laws in the form (9.27); I only needed I to be stationary.

- The choice of the q_i's is not restricted to Cartesian coordinates.

Example 9.4   Consider a free particle with Lagrangian:

$$L = \frac{1}{2} m \dot{\boldsymbol{r}}^2, \tag{9.28}$$

with

$$\boldsymbol{r} = (x, y, z).$$

Now measure the motion of the particle w.r.t. a rotating coordinate system with angular velocity

$$\boldsymbol{\omega} = (0, 0, \omega).$$

If $\mathbf{r}' = (x', y', z')$ are the coordinates in the rotating system, then

$$
\begin{aligned}
z &= z', \\
x &= x' \cos \omega t - y' \sin \omega t, \\
y &= y' \cos \omega t + x' \sin \omega t, \\
\Rightarrow \dot{z} &= \dot{z}', \\
\dot{x} &= \dot{x}' \cos \omega t - \dot{y}' \sin \omega t - x' \omega \sin \omega t - \omega y' \cos \omega t, \\
\dot{y} &= \dot{y}' \cos \omega t + \dot{x}' \sin \omega t - y' \omega \sin \omega t + x' \omega \cos \omega t, \\
\Rightarrow \dot{x}^2 + \dot{y}^2 + \dot{z}^2 &= \omega^2 [x'^2 + y'^2] + (\dot{x}'^2 + \dot{y}'^2) + 2\omega (x' \dot{y}' - \dot{x}' y') + \dot{z}'^2, \\
\Rightarrow L(\mathbf{r}', \dot{\mathbf{r}}') &= \frac{m}{2} \left[\omega^2 (x'^2 + y'^2) + (\dot{x}'^2 + \dot{y}'^2) + 2\omega (x' \dot{y}' - \dot{x}' y') \right].
\end{aligned}
$$

$$(9.29)$$

$$
\begin{aligned}
L(\mathbf{r}', \dot{\mathbf{r}}') &= \frac{m}{2} \left[\omega^2 (x'^2 + y'^2) + (\dot{x}'^2 + \dot{y}'^2) + 2\omega (x' \dot{y}' - \dot{x}' y') \right], \\
\frac{d}{dt} \left(\frac{\partial L}{\partial \dot{x}'} \right) &= \frac{d}{dt} (m \dot{x}' - \omega y') = m \ddot{x}' - \omega \dot{y}', \\
\frac{d}{dt} \left(\frac{\partial L}{\partial \dot{y}'} \right) &= \frac{d}{dt} (m \dot{y}' + \omega x') = m \ddot{y}' + \omega \dot{x}', \\
\frac{\partial L}{\partial x'} &= m \omega^2 x' + \omega \dot{y}', \\
\frac{\partial L}{\partial y'} &= m \omega^2 y' - \omega \dot{x}'.
\end{aligned}
$$

$$(9.30)$$

Then Euler–Lagrange gives us

$$
\begin{aligned}
\ddot{x}' &= \omega^2 x' + 2\omega \dot{y}', \\
\ddot{y}' &= \omega^2 y' - 2\omega \dot{x}', \\
\Rightarrow m \ddot{\mathbf{r}}' &= -m\boldsymbol{\omega} \times (\boldsymbol{\omega} \times \mathbf{r}') - 2m\boldsymbol{\omega} \times \dot{\mathbf{r}}'.
\end{aligned}
$$

$$(9.31)$$

Thus, we have recovered the "*fictitious forces*" characteristic of a non-inertial frame, the "*centrifugal*" and "*Coriolis*" terms [17].

9.3   CONSTRAINED VARIATIONS

We can employ the Lagrange multiplier method of Chapter 2 to incorporate constraints into the calculus of variations.

Suppose we want to find the path $y(x)$ which makes

$$
I = \int_{x_a}^{x_b} F(x, y, y') dx
$$

$$(9.32)$$

stationary, subject to the constraint that

$$J = \int_{x_a}^{x_b} G(x, y, y')dx - C = 0,$$ (9.33)

where C is a constant.

We generalize our earlier argument and introduce a two-parameter family of curves $Y(x, \epsilon_1, \epsilon_2)$

$$\begin{aligned} Y(x, \epsilon_1, \epsilon_2) &= y(x) + \epsilon_1 \eta_1(x) + \epsilon_2 \eta_2(x) \\ \eta_i(x_a) &= \eta_i(x_b) = 0, \quad i = 1, 2 \end{aligned}$$ (9.34)

η_i as differentiable as we need. Now

$$\begin{aligned} I(\epsilon_1, \epsilon_2) &= \int_{x_a}^{x_b} F(x, Y, Y')dx, \\ J(\epsilon_1, \epsilon_2) &= \int_{x_a}^{x_b} G(x, Y, Y')dx - C \end{aligned}$$ (9.35)

are two real-valued functions of ϵ_1 and ϵ_2. Define

$$L(\epsilon_1, \epsilon_2, \lambda) = I(\epsilon_1, \epsilon_2) - \lambda J(\epsilon_1, \epsilon_2),$$

λ is our Lagrange multiplier.

The conditions for an extremum are

$$\begin{aligned} \left[\frac{\partial L}{\partial \epsilon_i}\right]_{\epsilon_i = 0} &= 0 \quad i = 1, 2, \\ \left[\frac{\partial L}{\partial \lambda}\right]_{\epsilon_i = 0} &= 0. \end{aligned}$$ (9.36)

As before, the second equation is just the constraint. We can write

$$\begin{aligned} L(\epsilon_1, \epsilon_2, \lambda) &= \int_{x_a}^{x_b} H(Y, Y', x), \\ H &= F - \lambda G. \end{aligned}$$ (9.37)

Expanding in a Taylor series, (2.6), in ϵ_1 and ϵ_2 and integrating by parts then putting $\epsilon_i = 0$ we find that

$$\begin{aligned} \frac{\partial L}{\partial \epsilon_i} &= \int_{x_a}^{x_b} \left[\frac{\partial H}{\partial y} - \frac{d}{dx}\left(\frac{\partial H}{\partial y'}\right)\right] \eta_i(x)dx = 0, \\ &\Rightarrow \frac{\partial H}{\partial y} - \frac{d}{dx}\left(\frac{\partial H}{\partial y'}\right) = 0. \end{aligned}$$ (9.38)

This just like the Euler–Lagrange equation (9.20) except that $H = F - \lambda G$ replaces F. Note the solution of the Euler–Lagrange equation involves two constants of integration plus the constraint condition is enough to ensure that $y(x)$ passes through (x_a, a) and (x_b, b).

Generalization of these results to include multiple constraints is not difficult. If we have M constraints:

$$J_i = \int_{x_a}^{x_b} G_i(y, y', x)dx,$$
$$1 \leq i \leq M.$$

Define the function

$$H = F + \sum_{i=1}^{M} \lambda_i G_i \tag{9.39}$$

and look for the extrema of

$$L(y, y', \lambda_1, \ldots, \lambda_M) \equiv \int_{x_a}^{x_b} H(y, y', \lambda_1, \ldots, \lambda_M, x). \tag{9.40}$$

Then the new H defined in (9.40) satisfies the Euler–Lagrange equations.

Example 9.5   Let us look for the wave function $\psi(x)$ which minimizes

$$\langle \psi | \hat{H} \psi \rangle = \int_{-\infty}^{\infty} \bar{\psi}(x) \hat{H} \psi(x) dx. \tag{9.41}$$

Subject to the constraint that

$$\int_{-\infty}^{\infty} \bar{\psi}(x) \psi(x) dx = 1, \tag{9.42}$$

where

$$\hat{H} = -\frac{\hbar^2}{2m} \frac{d^2}{dx^2} + V(x). \tag{9.43}$$

We require

$$\lim_{x \to \pm\infty} \psi(x) = 0.$$

Therefore,

$$\int_{-\infty}^{\infty} \bar{\psi} \frac{d^2 \psi}{dx^2} dx = \bar{\psi} \frac{d\psi}{dx} \Big|_{-\infty}^{\infty} - \int_{-\infty}^{\infty} \frac{d\bar{\psi}}{dx} \frac{d\psi}{dx} dx. \tag{9.44}$$

Our task is to minimize

$$\langle \psi | \hat{H} \psi \rangle = \int_{-\infty}^{\infty} \left[\frac{\hbar^2}{2m} \left(\frac{d\bar{\psi}}{dx} \frac{d\psi}{dx} \right) + V \bar{\psi} \psi \right] dx \tag{9.45}$$

subject to (9.42). If we treat ψ and $\bar{\psi}$ as two dependent variables and add the constraint using a Lagrange multiplier ϵ the Euler–Lagrange equations become

$$V\psi - \epsilon\psi - \frac{\hbar^2}{2m} \left(\frac{d^2\psi}{dx^2} \right) = 0,$$
$$V\bar{\psi} - \epsilon\bar{\psi} - \frac{\hbar^2}{2m} \left(\frac{d^2\bar{\psi}}{dx^2} \right) = 0,$$
$$\int_{-\infty}^{\infty} \bar{\psi}(x)\psi(x)dx = 1. \tag{9.46}$$

The first and second of these equations in (9.46) are equivalent so we deduce that ψ satisfies the Schrödinger equation:

$$\left[-\frac{\hbar^2}{2m} \frac{d^2\psi}{dx^2} + V(x) \right] = \epsilon\psi. \tag{9.47}$$

We can thus identify the Lagrange multiplier with the energy of the physical system. It is a straightforward matter to extend this result to three dimension, i.e., the square integrable function $\psi(\boldsymbol{r})$ which minimizes

$$\langle \psi | \hat{H} \psi \rangle = \int \int \int \bar{\psi}(\boldsymbol{r}) \left[-\frac{\hbar^2}{2m} \nabla^2 \psi + V(\boldsymbol{r}) \right] \psi(\boldsymbol{r}) d^3\boldsymbol{r} \tag{9.48}$$

subject to the constraint that

$$\int \int \int \bar{\psi}(\boldsymbol{r})\psi(\boldsymbol{r})d^3\boldsymbol{r} = 1, \tag{9.49}$$

must satisfy the time independent Schrödinger equation:

$$\left[\frac{-\hbar^2}{2m} \nabla^2 \psi + V(\boldsymbol{r}) \right] \psi = \epsilon\psi. \tag{9.50}$$

9.4   STURM–LIOUVILLE REVISITED

Suppose we wanted to find the functions $y(x)$ for which

$$I[y] = \int_a^b (py(y')^2 + qy^2)dx \tag{9.51}$$

is stationary subject to the constraint that

$$G[y] = \int_a^b w(x)y^2 dx = 1 \tag{9.52}$$

$p(x), w(x) > 0$ on $[a, b]$. Let us introduce a Lagrange multiplier λ then (9.36) tells us that $I - \lambda G$ is stationary when

$$\frac{d2py'}{dx} = 2qy - 2\lambda wy$$

$$\Rightarrow -\frac{dpy'}{dx} + qy = \lambda wy \tag{9.53}$$

which is the Sturm–Liouville equation. If we multiply (9.53) by y and integrate we find

$$\begin{aligned}
\int_a^b \left[-y\frac{dpy'}{dx} + qy^2 \right] dx &= \lambda \int_{x_a}^{x_b} wy^2 dx \\
&= \lambda G[y] \\
&= \lambda . \tag{9.54}
\end{aligned}$$

It follows, integrating the first integral by parts, that

$$\begin{aligned}
\lambda &= \int_a^b \left[-y\frac{dpy'}{dx} + qy^2 \right] dx \\
&= -ypy'|_a^b + \int_a^b (py(y')^2 + qy^2)dx \\
&= I[y], \tag{9.55}
\end{aligned}$$

where we have assumed the usual Sturm–Liouville boundary conditions, (7.2). Thus, the stationary values of

$$\begin{aligned}
I[y] = &= -\frac{\int_a^b y(p(y')^2 + qy)dx}{\int_{x_a}^{x_b} y^2(x)w(x)dx} \\
&= \frac{\int_a^b y(py' - qy)dx}{\int_a^b y^2(x)w(x)dx} \tag{9.56}
\end{aligned}$$

are given by

$$F[y_n(x)] = \lambda_n,$$

where λ_n are the eigenvalues of the Sturm–Liouville operator corresponding to the eigenfunctions y_n.

In summary, the following three problems are equivalent:

(i) Find the eigenvalues, λ and the eigenfunctions $y(x)$ that solve the Sturm–Liouville problem

$$-\frac{d(p(x)y')}{dx} + q(x)y = \lambda w(x)y,$$

$$-ypy'\Big|_a^b = 0,$$

$$w(x), p(x) > 0, x \in (a,b).$$

(ii) Find the function $y(x)$ for which

$$F[y] = \int_a^b \left(py'^2 + qy^2\right) dx$$

is stationary subject to the constraint that

$$G[y] = \int_a^b wy^2 dx = 1.$$

The eigenvalues of the equivalent Sturm–Liouville problem in (i) are given by $F[y]$.

(iii) Find the function $y(x)$ for which

$$H[y] = \frac{F[y]}{G[y]}$$

is stationary. The eigenvalues of the Sturm–Liouville problem are then given by the values of $H[y]$.

We can make use of these equivalences to estimate the eigenvalues and eigenfunctions of Sturm–Liouville problem.

Example 9.6   Consider the simple Sturm–Liouville problem

$$u'' + \lambda u = 0,$$
$$u(0) = u(1) = 0, \tag{9.57}$$

whose solutions we know to be

$$u_n = \sin(n\pi x)$$
$$\lambda_n = (n\pi)^2$$
$$n = 1, 2, \ldots. \tag{9.58}$$

We are looking for the lowest eigenvalue. Let us try two different test functions.

(i) The "*hat*" function

$$u_h(x) = \begin{cases} x, & 0 \le x \le \frac{1}{2} \\ 1 - x, & \frac{1}{2} \le x \le 1. \end{cases} \tag{9.59}$$

In our case, $w(x) = 1.0$. Hence,

$$\begin{aligned} \int_0^1 u_h^2 w(x) dx &= \int_0^{\frac{1}{2}} x^2 dx + \int_{\frac{1}{2}}^1 [1 - 2x + x^2] dx, \\ &= 2 \frac{x^3}{3} \Big|_0^{\frac{1}{2}}, \\ &= \frac{1}{12}. \end{aligned} \tag{9.60}$$

We can normalize our trial function $\bar{u}_h = \sqrt{12} u_h$.

(ii) The quadratic trial function

$$\begin{aligned} u_q &= x(x - 1), \\ \int_0^1 u_q^2 dx &= \frac{1}{30}, \\ \bar{u}_q &= \sqrt{30} u_q. \end{aligned} \tag{9.61}$$

In Figure 9.2, I show a comparison between the normalized trial functions $\bar{u}_h$ and $\bar{u}_q$ with the exact solution $u_1(x)$. Now let us consider the Rayleigh quotient for both trial functions to get an estimate for the eigenvalue λ:

$$\begin{aligned} \lambda_h &= \frac{\int_0^1 (u_h')^2 dx}{(\int_0^1 u_h^2 w(x) dx)^2)} &= 12, \\ \lambda_q &= \frac{\int_0^1 (u_q')^2 dx}{(\int_0^1 u_q^2 w(x) dx)^2)} &= 10. \end{aligned} \tag{9.62}$$

The true value of the eigenvalue is 9.8626, as expected both approximate values are greater than the exact value, with λ_h overestimating the exact value by approximately 21% and λ_q over estimating it by only 1.3%.

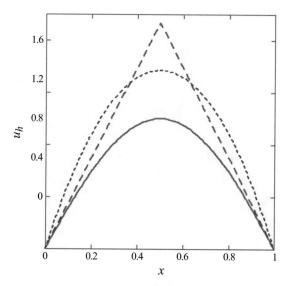

Figure 9.2: Trial functions, $u_h(x)$, red long dashed, $u_q(x)$, blue short dashed compared with exact eigenfunction $u_1(x)$, solid green.

CHAPTER 10

Case Study: The Ground State of Atoms

In nature every undisturbed quantum system tends to its lowest allowed energy. This state is known as the ground state. An excited state is any state with energy greater than the ground state. It is not possible to make a direct measurement of the quantum wave function. A quality measure of an approximate wave function is how well it reproduces the ground state energy. In this chapter I plan to focus on the computation of the ground state energies of multi-electron atoms/ions with special reference to the application of the variational methods developed in the previous chapter.

The Schrödinger equation for a one electron ion admits an analytic solution [23] which we can use as a staring point for treating the more complex systems. (Atomic units with $\hbar = 1, e = 1, m_e = 1, a_0 = 1, 4\pi\epsilon_0 = 1$ are used throughout this chapter.)

10.1 HYDROGENIC IONS

The energies of the stationary states of a hydrogenic ion of charge Z are the eigenvalues of the time-independent Schrödinger equation:

$$\left[-\frac{1}{2\mu}\nabla^2 - \frac{Z}{r} \right] \psi(\mathbf{r}) = E\psi(\mathbf{r}). \tag{10.1}$$

μ is the reduced mass:

$$\frac{M}{M + 1},$$

where M is the nuclear mass. The eigenfunctions $\psi_{nlm}(\mathbf{r})$ can be written [23]

$$\psi_{nlm}(\mathbf{r}) = R_{nl}(r)Y_l^m(\theta, \phi), \tag{10.2}$$

where n, l, m are integers, s.t.

$$\begin{aligned} n &= 1, 2, \ldots, \\ l &= 0, 1, \ldots, n - 1, \\ m &= -l, -l + 1, \ldots, l - 1, l \end{aligned}$$

and $Y_l^m(\theta, \phi)$ is a "*spherical harmonic.*"

The first few radial functions are given by [23, 26]

$$
\begin{aligned}
R_{10}(r) &= 2(Z\mu)^{3/2}\exp(-Z\mu r), \\
R_{20}(r) &= 2(\tfrac{1}{2}Z\mu)^{3/2}\left(1 - \frac{Z\mu r}{2}\right)\exp(-\tfrac{1}{2}Z\mu r), \\
R_{21}(r) &= \frac{1}{\sqrt{3}}(\tfrac{1}{2}Z\mu)^{3/2}(Z\mu r)\exp(-\tfrac{1}{2}Z\mu r), \\
R_{30}(r) &= 2(\frac{Z\mu}{3})^{3/2}\left(1 - 2\frac{Z\mu r}{3} + 2\frac{(Z\mu)^2 r^2}{27}\right)\exp(-\frac{Z\mu r}{3}), \\
R_{31}(r) &= \frac{4\sqrt{2}}{9}(\frac{Z\mu}{3})^{3/2}\left(1 - \frac{Z\mu r}{6}\right)(Z\mu r)\exp(-\frac{Z\mu r}{3}), \\
R_{32}(r) &= \frac{4}{27\sqrt{10}}(\frac{Z\mu}{3})^{3/2}(Z\mu r)^2\exp(-\frac{Z\mu r}{3})
\end{aligned}
\tag{10.3}
$$

and the energy eigenvalues are given by

$$
E_n = -\frac{Z\mu}{2n^2} = -\frac{Z\mu}{n^2}Ryd,
\tag{10.4}
$$

where Ryd is the Rydberg energy which is just $\frac{1}{2}$ in atomic units. We are interested in the ground state here:

$$
\psi_{1,0,0} = R_{1,0}(r)Y_0^0(\theta, \phi) = 2(Z\mu)^{3/2}\exp(-Z\mu r)\left(\frac{1}{4\pi}\right)^{\frac{1}{2}}.
\tag{10.5}
$$

Since the nucler mass it is very much bigger than the electron mass we can just take $\mu = 1$ which is what I will do from now on.

10.2   TWO ELECTRON IONS

The Hamiltonìan for two electrons, each of charge -1, orbiting a nucleus of charge Z is

$$
H = -\frac{1}{2}\nabla_1^2 - \frac{Z}{r_1} - \frac{1}{2}\nabla_1^2 - \frac{Z}{r_2} + \frac{1}{||r_1 - r_2||}.
\tag{10.6}
$$

For helium $Z = 2$ but it will be convenient to keep Z arbitrary for the time being. If I make the orbital approximation [23] and ignore the electron-electron interaction the problem is separable and the eigenstate for the two electron system are of the form:

$$
\Psi_{n_1,l_1,m_1,n_2,l_2,m_2}(r_1, r_2) = \psi_{n_1,l_1,m_1}(r_1)\psi_{n_2,l_2,m_2}(r_2),
\tag{10.7}
$$

where $\psi_{n_i l_i m_i}(r_i)$ are the usual energy eigenstates of the hydrogenic ion with nuclear charge Z. You should remember that the electrons are fermions so we cannot put them in the same state.

However, electrons also have a spin degree of freedom which we have neglected in (10.7). This means that two electrons can have the same spatial wavefunction as long as one is spin up and the other spin down. The interaction energy in this approximation is just the sum of the energies of both orbitals:

$$E = -Z^2 \left[\frac{1}{n_1^2} + \frac{1}{n_2^2} \right] Ryd.$$

Setting $Z = 2, n_1 = n_2 = 1$ for helium we get a ground state energy of $-8Ryd \approx -108.8$ eV. The ground state of helium has a measured energy which is very close to -79 eV. Clearly, we need to take into account the interaction term to get a better estimate. I will explore two methods to finding an improved approximate solution for the two electron ion.

- I will look at a "*perturbative*" approach. Perturbation theory is a systematic method for finding an approximate solution to a problem, by starting from the exact solution of a related, simpler problem. I will only use the first-order theory as outlined in Appendix C.

- The variational approach discussed in the previous chapter.

We can take $\Psi_{n_1,l_1,m_1,n_2,l_2,m_2}(r_1, r_2)$ as defined in (10.7) to be our unperturbed state. If we are to apply perturbation theory we need to be able to assume that the neglected term, the $e^- e^-$ interaction is smaller than the unperturbed term. Both the electron-nucleus and electron-electron terms are Coulomb interactions differing by a factor Z so crudely our perturbation is $\frac{1}{Z}$ smaller. This is only a half for helium so we might expect that perturbation theory will only give a very crude estimate of the correction. Our hydrogenic ground state orbital is given by (10.5):

$$\psi_{1,0,0}(r) = (Z)^{3/2} \exp(-Zr) \frac{1}{\sqrt{\pi}}.$$

Hence,

$$
\begin{aligned}
\Delta E &= \langle \Psi_{1,0,0,1,0,0} | H_I \Psi_{1,0,0,1,0,0} \rangle \\
&= \int d^3 r_1 d^3 r_2 \frac{|\psi_{1,0,0}(r_1)|^2 |\psi_{1,0,0}(r_2)|^2}{||r_1 - r_2||} \\
&= \frac{5Z}{4} Ryd,
\end{aligned}
\tag{10.8}
$$

where I have made us of the integral [23]

$$\int d^3 r_1 d^3 r_2 \frac{e^{-\gamma[r_1 + r_2]}}{||r_1 - r_2||} = \frac{20\pi^2}{\gamma^5}. \tag{10.9}$$

So, the first-order correction is a positive term and yields a ground state energy:

$$\left(-8 + \frac{5}{2}\right) Ryd \approx -74.8 \text{ eV}. \tag{10.10}$$

This is not a bad first estimate. However to take the perturbation to higher orders is very demanding.

Now let us try the variational approach, taking as our variational test function a normalized wave function:

$$\psi_t(\mathbf{r}_1, \mathbf{r}_2) = \frac{\tilde{Z}^3}{\pi} e^{-\tilde{Z}(r_1+r_2)}. \tag{10.11}$$

Our trial function looks like the product of two hydrogenic functions for a nuclear charge $\tilde{Z}$ but this "*charge*" is not a real constant charge but a variable parameter which we can chose at will in order to make use of the Rayleigh–Ritz theorem.

$$
\begin{aligned}
\langle \psi_t | \hat{H} \psi_t \rangle &\equiv \langle \hat{H} \rangle \\
&= \int d^3 r_1 d^3 r_2 \bar{\psi}_t \\
&\quad \times \left(-\frac{1}{2}\nabla_1^2 - \frac{1}{2}\nabla_2^2 - [\frac{\tilde{Z}}{r_1} + \frac{\tilde{Z}}{r_2}] + \frac{\tilde{Z} - Z}{r_1} + \frac{\tilde{Z} - Z}{r_2} + \frac{1}{\|\mathbf{r}_1 - \mathbf{r}_2\|} \right) \psi_t.
\end{aligned}
\tag{10.12}
$$

Now since we are using hydrogenic functions, it can be shown [23, 26],

$$
\begin{aligned}
\langle \psi_t | \left(-\frac{1}{2}\nabla_1^2 - \frac{1}{2}\nabla_2^2 - \left[\frac{\tilde{Z}}{r_1} + \frac{\tilde{Z}}{r_2} \right] \right) \psi_t \rangle &= -\tilde{Z}^2, \\
\langle \psi_t | \frac{\tilde{Z} - Z}{r_1} + \frac{\tilde{Z} - Z}{r_2} \psi_t \rangle &= 2(\tilde{Z} - Z)\langle \frac{1}{r} \rangle \\
\langle \psi_t | \frac{1}{\|\mathbf{r}_1 - \mathbf{r}_2\|} \psi_t \rangle &= -\frac{5\tilde{Z}}{8},
\end{aligned}
\tag{10.13}
$$

where $\langle \frac{1}{r} \rangle$ is the expectation for the ground state for a one electron hydrogenic ion with nuclear charge $\tilde{Z}$ and is equal to $\tilde{Z}$. With everything in Rydbergs we have

$$\langle \hat{H} \rangle = [-2\tilde{Z}^2 + 4\tilde{Z}(\tilde{Z} - Z) + \frac{5}{4}\tilde{Z}] \; Ryd. \tag{10.14}$$

Let us now find the value of $\tilde{Z}$ which gives the minimum value, $\tilde{Z}_{\min}$:

$$
\begin{aligned}
\frac{d\langle \hat{H} \rangle}{d\tilde{Z}} \bigg|_{\tilde{Z}_{\min}} &= 0, \\
\Rightarrow -4\tilde{Z}_{\min} + 8\tilde{Z}_{\min} - 4Z + \frac{5}{4} &= 0, \\
\Rightarrow \tilde{Z}_{\min} &= Z - \frac{5}{16}.
\end{aligned}
\tag{10.15}
$$

For helium $Z = 2$ and consequently $\tilde{Z}_{min} = 27/16$ and our upper bound on the ground state energy is

$$\langle \hat{H} \rangle = - \left(\frac{27}{16} \right)^2 Ryd \approx -77.46 \text{ eV}. \tag{10.16}$$

One of the advantages of the variational approach is not only does it give an estimate of the ground state but also an approximate wave function. In this case we can interpret our new wave function by saying that each electron moves on average in the field of a nucleus with charge $\tilde{Z}$, rather than charge Z. The difference between $\tilde{Z}$ and Z is a measure of the degree of screening due to the second electron.

So far, we have not fixed Z so we can apply the same analysis to other two electron systems. The negative ion of hydrogen H^- is an interesting testing ground for exploring variational techniques for estimating atomic wavefunctions [27, 28]. If we use the test function (10.11) our analysis follows through exactly as before but now $Z = 1$ and $\tilde{Z}_{min} = 11/16$ we thus have an upper bound on the energy of the three particle system

$$- \left(\frac{11}{16} \right)^2 Ryd \approx -12.86 \text{ eV}. \tag{10.17}$$

Now this energy is greater than the ground state energy of neutral hydrogen (-13.6 eV). Thus, if this were the actual energy of the H^- ground state then it would be more energetically favorable to free one electron and the other electron to be left in the ground state of the neutral. The variational method only gives us an upper bound on the energy so it would be premature to assume that there are no bound states of H^- based on this calculation alone. Bethe [29] used a trial function which depended on a three parameter function of the form:

$$\psi = (1 + \alpha u + \beta t^2)e^{-\kappa s}$$
$$\text{where}$$
$$u \equiv ||r_1 - r_2||,$$
$$s \equiv r_1 + r_2,$$
$$t \equiv r_1 - r_2, \tag{10.18}$$

and α, β, κ are the variational parameters. It was shown that with this wave function the resulting Rayleigh–Ritz upper bound on the energy lies below $-1Ryd$. More and more sophisticated and complex trial functions have been used. The best current estimate ground energy is close to -14.36 eV.

H^- is of astrophysical importance [27]. The abundant presence of both hydrogen and low energy electrons in the ionized atmospheres of the Sun and other stars is ideal for the creation of H^- by electron attachment. Radiation from the surface of the sun is absorbed by photodetachment. The continual formation and destruction of the negative ion conserves the total radiated energy but modifies the characteristics of the light emitted from the star. Indeed, since most neutral atoms and positive ions have their first absorption at 4 or 5 eV if not larger, H^- is the dominant contributor to the absorption of 0.75 eV photons, a critical range of infrared and visible wavelengths.

Chandrasekar [30] used a two-parameter trial function:

$$\psi_{trial}(r_1, r_2) = \frac{N}{4\pi}\left[e^{-\tilde{Z}_1 r_1 - \tilde{Z}_2 r_2} + e^{-\tilde{Z}_1 r_2 - \tilde{Z}_2 r_1}\right]. \tag{10.19}$$

Notice we have two variational parameters $\tilde{Z}_1$ and $\tilde{Z}_2$ and that our wave function is symmetric under the interchange of r_1 and r_2. Using (10.19) and Rayleigh–Ritz Chandrasekar found $\tilde{Z}_1 = 1.039$ and $\tilde{Z}_2 = .283$ and an upper bound on the ground state energy of -13.98, slightly less than the binding energy of hydrogen and not too far of the actual ground state energy. The function exhibits a "*radial correlation*" only. Particularly striking is the feature that $\tilde{Z}_1$ is larger than 1, we can interpret this as implying that the effect of the second electron is to force the inner one closer to the nucleus than it would be were it alone bound to the proton. The more complex wave functions like those of the Bethe, (10.18) include "*angular correlation*" between the directions $\hat{r}_1$ and $\hat{r}_2$ and "*radial*" between the magnitudes r_1 and r_2, the fact that the Chandrasekar wavefunction gives a "*good*" bound state energy suggests that radial is the more significant of the two types of correlation.

10.3   THE HARTREE APPROACH

Let us consider an atom with N electrons. We are looking for an optimal trial wavefunction to use in our variational theorem. As a first approximation I will ignore spin and not impose antisymmetry on the wave function. As my test function I will take

$$\Psi(r_1, r_2, \ldots, r_N) = \psi_{\alpha_1}(r_1)\psi_{\alpha_2}(r_2)\ldots\psi_{\alpha_N}(r_N), \tag{10.20}$$

where I assume each one particle function is normalized and can be characterized by a set of quantum numbers α. $\psi_\alpha(r_q)$ is the one particle wave function for electron q with the set of quantum numbers α. These quantum numbers could be $\alpha \equiv n_i, l_i, m_i$. We do not have to assume the single particle functions are hydrogenic wave functions. In practical calculations Slater-type orbitals or Gaussian orbitals are frequently used [31].

The full multi-electron Hamiltonian is:

$$\hat{H} = \sum_{i=1}^{N}\left[\left(-\frac{1}{2}\nabla_i^2 - \frac{Z}{r_i}\right) + \sum_{j>i}\frac{1}{r_{ij}}\right]. \tag{10.21}$$

Now using Ψ as our trial function the expectation value of the energy becomes:

$$\langle\Psi|\hat{H}\psi\rangle \equiv \langle\hat{H}\rangle$$
$$= \sum_{i=1}^{N}\left[\int d^3r\,\bar{\psi}_{\alpha_i}(r)\left(-\frac{1}{2}\nabla^2 - \frac{Z}{r}\right)\psi_{\alpha_i}(r)\right.$$

$$+ \sum_{j>i} \int d^3r d^3r' \frac{\bar{\psi}_{\alpha_i}(\boldsymbol{r})\bar{\psi}_{\alpha_j}(\boldsymbol{r}')\psi_{\alpha_i}(\boldsymbol{r})\psi_{\alpha_j}(\boldsymbol{r}')}{\|\boldsymbol{r}-\boldsymbol{r}'\|} \Bigg]. \qquad (10.22)$$

Let us focus for a moment on the second term. Let

$$J_{ij} = \int d^3r d^3r' \frac{\bar{\psi}_{\alpha_i}(\boldsymbol{r})\bar{\psi}_{\alpha_j}(\boldsymbol{r}')\psi_{\alpha_i}(\boldsymbol{r})\psi_{\alpha_j}(\boldsymbol{r}')}{\|\boldsymbol{r}-\boldsymbol{r}'\|}. \qquad (10.23)$$

Since the integral is over $\boldsymbol{r}$ and $\boldsymbol{r}'$ we have that $J_{ij} = J_{ji}$, hence

$$\sum_{j>i} J_{ij} = \frac{1}{2} \sum_{j \neq i} J_{ij}, \qquad (10.24)$$

and we may write

$$\begin{aligned} \langle \hat{H} \rangle &= \sum_{i=1}^{N} \Bigg[\int d^3r \, \bar{\psi}_{\alpha_i}(\boldsymbol{r}) \left(-\frac{1}{2}\nabla^2 - \frac{Z}{r} \right) \psi_{\alpha_i}(\boldsymbol{r}) \\ &+ \frac{1}{2} \sum_{j \neq i} \int d^3r d^3r' \frac{\bar{\psi}_{\alpha_i}(\boldsymbol{r})\bar{\psi}_{\alpha_j}(\boldsymbol{r}')\psi_{\alpha_i}(\boldsymbol{r})\psi_{\alpha_j}(\boldsymbol{r}')}{\|\boldsymbol{r}-\boldsymbol{r}'\|} \Bigg]. \end{aligned} \qquad (10.25)$$

To find the least upper bound on the energy with this ansatz (10.20) we need to we minimize $\langle \hat{H} \rangle$ over all possible one particle orbitals. If we keep each orbital ψ_{α_i} normalized then the N particle wave function Ψ will be normalized. To achieve this we introduce N Lagrange multipliers, ϵ_i. Consider the functional

$$F[\Psi] = \langle \hat{H} \rangle - \sum_i \epsilon_i \left(\int d^3r |\psi_{\alpha_i}(\boldsymbol{r})|^2 - 1 \right). \qquad (10.26)$$

We want to find the wave functions ψ_{α_i} which will make F minimal. Just as in Example 9.5, we can vary its real and imaginary parts independently. Since we have N independent wavefunctions, this gives rise to $2N$ real conditions. This gives two sets of N complex equations, however one set is simply the conjugate of the other and so we only need the N equations:

$$\left[-\frac{1}{2}\nabla_i^2 - \frac{Z}{r_i} + \sum_{i \neq j} \int d^3r' \frac{\bar{\psi}_{\alpha_j}(\boldsymbol{r}')\psi_{\alpha_j}(\boldsymbol{r}')}{\|\boldsymbol{r}-\boldsymbol{r}'\|} \right] \psi_{\alpha_i}(\boldsymbol{r}) = \epsilon_i \psi_{\alpha_i}(\boldsymbol{r}), \qquad (10.27)$$

called the Hartree equations.

Notice that in taking the variation derivative with respect to $\bar{\psi}_n$ we end up with two identical terms corresponding to $n = i$ and $n = j$ and we lose the factor of a half in (10.25).

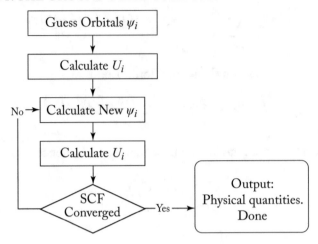

Figure 10.1: Flow chart for a self-consistent field computer code.

Equation (10.27) has the same form as the regular Schrödinger equation where we have an effective potential:

$$U_i(r) = \sum_{j \neq i} \int d^3 r' \frac{\bar{\psi}_{\alpha_j}(r')\psi_{\alpha_j}(r')}{\|r - r'\|}, \tag{10.28}$$

and our Lagrange multipliers are now the orbital energies. We interpret $U_i(r)$ as coming from the electrostatic potential due to all the electrons other than i. The thing to notice is that each ψ_{α_j} that appears in $U_i(r)$ is itself determined by one of the Hartree equations in (10.27) so we have a set of coupled integro-differential equations. The potentials U_i both determine the wavefunctions and are determined by the wavefunctions. The major requirement now is "*self-consistency.*" The usual way forward is to proceed iteratively, see Figure 10.1. We write down a physically reasonable guess for our product wavefunction (10.20) and use this to calculate U_i then calculate a new set of energies energies ϵ_i and orbitals ψ_{α_i} from which we can calculate a new U_i and continue like this until we have reached the desired level of convergence. Now taking the inner product of (10.27) with $\psi_{\alpha_i}(r)$ yields

$$\epsilon_i = \int d^3 r \bar{\psi}_{\alpha_i}(r)\left[-\frac{1}{2}\nabla_i^2 - \frac{Z}{r_i}\right]\psi_{\alpha_i}(r) + \sum_{i \neq j} \int d^3 r d^3 r' \frac{|\psi_{\alpha_j}(r')|^2 |\psi_{\alpha_i}(r)|^2}{\|r - r'\|}. \tag{10.29}$$

Summing over i we almost get the expression (10.22). Unfortunately, the inner summation in (10.22) is over $j > i$ while in (10.27) it is over $j \neq i$ which means, as pointed out above (10.24), it double counts. Correcting for the double counting means our variational estimate for the ground state is:

$$E_{variational} = \sum_i \left(\epsilon_i - \sum_{i<j} \int d^3r d^3r' \frac{|\psi_{\alpha_j}(r')|^2 |\psi_{\alpha_i}(r)|^2}{\|r - r'\|} \right). \tag{10.30}$$

The Hartree approach has been generalized to take account of spin in the "*Hartree–Fock*" theory, where the N-body test wave function of the system is taken to be antisymmetric products of one electron orbitals, a "*Slater determinant*." For more details, see [23, 24].

APPENDIX A

Vector Spaces

Vector space theory lies at the heart of much, if not most, numerical methods. We commonly utilize the theorems of finite dimensional linear algebra and profit from our knowledge of self-adjoint operators in infinite dimensional Hilbert spaces. In this appendix, I have gathered together some key results and observations that are employed throughout this book.

Definition A.1   A vector space, $\mathbb{V}$, over the complex numbers, $\mathbb{C}$, is a set, Ω, together with operations of addition and multiplication by complex numbers which satisfy the following axioms. Given any pair of vectors x, y in $\mathbb{V}$ there exists a unique vector $x + y$ in $\mathbb{V}$ called the sum of x and y. It is required that

-
$$x + (y + z) = (x + y) + z,$$

  i.e., we require addition to be associative.

-
$$x + y = y + x,$$

  i.e., we require addition to be commutative.

- There exists a vector 0 s.t.
$$x + 0 = x.$$

- For each vector x there exists a vector $-x$ s.t.
$$x + (-x) = 0.$$

  Given any vector x in $\mathbb{V}$ and any α, β in $\mathbb{C}$ there exists a vector αx in $\mathbb{V}$ called the product of x and α. It is required that

-
$$\alpha(y + z) = \alpha y + \alpha z.$$

•

$$(\alpha + \beta)x = \alpha x + \beta x.$$

•

$$(\alpha\beta)x = \alpha(\beta x).$$

•

$$(1)x = x.$$

Definition A.2   A vector x is said to be linearly dependent on vectors $x_1 \ldots x_N$, if x can be written as:

$$x = \alpha_1 x_1 + \alpha_2 x_2 + \cdots + \alpha_N x_N.$$

If no such relation exists, then x is said to be linearly independent of $x_1 \ldots x_N$.

Now every vector in $\mathbb{R}^3$ can be written in terms of the three unit vectors e_x, e_y, e_z; we can generalize this idea.

Definition A.3   Suppose $\mathbb{V}$ is a vector space and there exists a set of vectors $\{e_i\}_{i=1}^N$. Then this set spans $\mathbb{V}$ or equivalently forms a basis for it. If

• the set of vectors $\{e_i\}_{i=1}^N$ is linearly independent, and

• if $c \in \mathbb{V}$ then it can be written:

$$c = \sum_{i=1}^{N} \alpha_i e_i.$$

It is easy to verify that the set of ordered n-tuples, of real numbers

$$x = (x_1, \ldots, x_n)$$

form a vector space over $\mathbb{R}$ when define addition and multiplication by a scalar by

$$\alpha x + \beta y \equiv (\alpha x_1 + \beta y_1, \ldots, \alpha x_n + \beta y_n). \tag{A.1}$$

Our "*ordinary*" vectors in $\mathbb{R}^3$ are just a special case. In $\mathbb{R}^3$ we have a scalar product

$$x \cdot y \equiv \sum_{i=1}^{3} x_i y_i. \tag{A.2}$$

We can generalize this for an arbitrary vector space, $\mathbb{V}$ over the complex numbers.

Definition A.4   An inner product is a map which associates two vectors in the space, $\mathbb{V}$, with a complex number

$$\langle | \rangle \quad : \mathbb{V} \times \mathbb{V} \quad \to \mathbb{C},$$
$$: a, b \quad \mapsto \langle a|b \rangle$$

that satisfies the following four properties for all vectors $a, b, c \in \mathbb{V}$ and all scalars. $\alpha, \beta \in \mathbb{C}$:

$$
\begin{aligned}
\langle a|b \rangle &= \overline{\langle b|a \rangle}, \\
\langle \alpha a|\beta b \rangle &= \overline{\alpha}\beta \langle a|b \rangle, \\
\langle a + b|c \rangle &= \langle a|c \rangle + \langle b|c \rangle, \\
\langle a|a \rangle &\geq 0 \text{ with equality iff } a = 0,
\end{aligned}
\tag{A.3}
$$

where $\overline{z}$ denotes the complex conjugate of z. We note that $\langle \alpha a|\alpha a \rangle = |\alpha|^2 \langle a|a \rangle$ which is consistent with the last property in (A.3).

We can now define the following.

Definition A.5   For a vector $a \in \mathbb{V}$ we can define its norm

$$\|a\| = \sqrt{\langle a|a \rangle},$$

it being understood that we take the positive square root.

We can generalize $\mathbb{R}^n$ to $\mathbb{C}^n$ where $r \in \mathbb{C}^n$ can be represented by the n-tuple of complex numbers

$$r = (z_1, \ldots, z_n).$$

Addition and multiplication by scalar will go through just as usual, i.e., if

$$
\begin{aligned}
r_1 &= (z_1, \ldots, z_n), \\
r_2 &= (\zeta_1, \ldots, \zeta_n),
\end{aligned}
$$

then

$$
\begin{aligned}
\alpha r_1 &= (\alpha z_1, \ldots, \alpha z_n), \\
r_1 + r_2 &= (z_1 + \zeta_1, \ldots, z_n + \zeta_n).
\end{aligned}
$$

However, if we are going to use our definition of inner product we will require:

$$\langle r_1|r_2 \rangle = \sum_{i=1}^{n} \overline{z}_i \zeta_i. \tag{A.4}$$

Consequently,

$$\|r_1\|^2 = \sum_{i=1}^{n} |z_i|^2.$$

We can deduce from Definition A.4 of the inner product.

Lemma A.6   *If $u, v \in \mathbb{V}$ then*

$$|\langle u|v \rangle| \le \|u\|\|v\|.$$

Proof. Let

$$w = u + \lambda v.$$

Then,

$$\langle w|w \rangle \ge 0.$$

But

$$
\begin{aligned}
\langle w|w \rangle &= \langle u|u \rangle + \lambda \langle u|v \rangle + \overline{\lambda} \langle v|u \rangle + |\lambda|^2 \langle v|v \rangle \\
&= \|u\|^2 + \lambda \langle u|v \rangle + \overline{\lambda} \langle v|u \rangle + |\lambda|^2 \|v\|^2 \ge 0.
\end{aligned}
\tag{A.5}
$$

Take

$$\lambda = -\frac{\langle v|u \rangle}{\|v\|^2},$$

and using

$$\langle u|v \rangle = \overline{\langle v|u \rangle}$$

(A.5) becomes

$$\|u\|^2 - \frac{|\langle u|v \rangle|^2}{\|v\|^2} \ge 0.$$

$\square$

This result is known as the Cauchy–Schwarz inequality.

Definition A.7   Two vectors $a, b \in \mathbb{V}$ are said to be orthogonal if

$$\langle a|b \rangle = 0.$$

If further

$$\langle a|a \rangle = \langle b|b \rangle = 1,$$

the vectors are said to be orthonormal.

Lemma A.8   *If $\{a_i\}_{i=1}^N$ are a set of mutually orthogonal non-zero vectors then they are linearly independent.*

Proof. Suppose

$$\sum_{i=1}^N \alpha_i a_i \;=\; 0,$$

$$\Rightarrow \sum_{i=1}^N \alpha_i \langle a_q|a_i \rangle \;=\; 0,$$

$$\Rightarrow \alpha_q \langle a_q|a_q \rangle \;=\; 0,$$

$$\Rightarrow \alpha_q \;=\; 0.$$

This is enough to establish linear independence.   □

Lemma A.9   *If $\{a_n\}_{i=1}^N$ is a set of linearly independent vectors which span $\mathbb{V}$ then there exists an orthonormal set $\{e_n\}_{i=1}^N$ which also spans $\mathbb{V}$.*

Proof. I will prove the result by explicitly constructing a set $\{e_n\}_{i=1}^N$

$$e_1 \;=\; \frac{a_1}{\|a_1\|},$$

$$e_2' \;=\; a_2 - \langle a_2|e_1 \rangle e_1,$$

$$e_2 \;=\; \frac{e_2'}{\|e_2'\|},$$

$$\vdots$$

$$e_N' \;=\; a_N - \sum_{k=1}^{N-1} \langle a_N|e_k \rangle e_k,$$

$$e_N \;=\; \frac{e_N'}{\|e_N'\|}. \tag{A.6}$$

□

The method of creating an orthonormal basis I just employed is known as the Grahm–Schmidt orthogonalization method.

Lemma A.10   *Let $\mathbb{V}$ be a vector space over $\mathbb{C}$ and let $\{e_1, \ldots, e_N\}$ be a basis for $\mathbb{V}$. Let $\{w_i\}_{i=1}^{M}$ be a set of non-zero vectors in $\mathbb{V}$. If $M > N$ then the set $\{w_i\}$ is linearly dependent.*

Proof. Let us begin by assuming that, on the contrary, the set $\{w_i\}$ is linearly independent. Since $\{e_i\}$ forms a basis we may write

$$w_1 = \sum_{i=1}^{N} \alpha_i e_i,$$

at least one α_i must be non-zero, renumbering if necessary we can chose it to be α_1

$$\Rightarrow e_1 = \alpha_1^{-1}[w_1 + \sum_{i=2}^{N} \alpha_i e_i].$$

Hence, the set $\{w_1, e_2, \ldots e_N\}$ spans $\mathbb{V}$.

Now we may repeat the argument for w_2,

$$w_2 = \beta_1 w_1 + \sum_{i=2}^{N} \beta_i e_i,$$

and since we are assuming $\{w_i\}$ is linearly independent, then at least one β_i with $i \geq 2$ is non-zero. We can keep repeating the argument until $w_1, \ldots, w_N$ spans $\mathbb{V}$ then since w_{N+1} an element of the space it can be written:

$$w_{N+1} = \sum_{i=1}^{N} \alpha_i w_i,$$

thus linearly dependent and we have a contradiction. Our original assumption is false and the result is established. $\qquad\square$

LINEAR OPERATORS AND MATRICES

Definition A.11   A linear operator $\hat{T}$ is a map from a vector space $\mathbb{V}$ onto itself s.t. for all $x, y \in \mathbb{V}, \alpha, \beta \in \mathbb{C}$

$$\hat{T}(\alpha x + \beta y) = \alpha \hat{T}(x) + \beta \hat{T}(y).$$

Now,

$$r = a + \tau b$$

defines the equation of a line through a parallel to the vector b [11]; then

$$\hat{T}[r] = \hat{T}[a] + \tau \hat{T}[b].$$

This is the equation of a line through $\hat{T}[a]$ parallel to $\hat{T}[b]$.

The linear operator $\hat{T}$ maps the vector space onto itself consequently if $\{e_i\}_{i=1}^{N}$ is an orthonormal basis for the finite dimensional space $\mathbb{V}$ then for each e_i we must be able to expand $\hat{T}(e_i)$ in terms of the full basis

$$\hat{T}(e_i) = \sum_{j=1}^{N} T_{ji} e_j, \tag{A.7}$$

where T_{ji} are complex numbers, taking the inner product with e_q we have

$$\begin{aligned}
\langle e_q | \hat{T}(e_i) \rangle &= \sum_{i=1}^{N} T_{ji} \langle e_q | e_j \rangle \\
&= \sum_{i=1}^{N} T_{ji} \delta_{qj} = T_{qi}.
\end{aligned} \tag{A.8}$$

Then if r is any vector in $\mathbb{V}$ we may expand it in terms of the basis:

$$\begin{aligned}
r &= \sum_{i=1}^{N} x_i e_i, \\
\Rightarrow \hat{T}(r) &= \sum_{i=1}^{N} x_i \hat{T}(e_i) \\
&= \sum_{i=1}^{N} x_i \sum_{j=1}^{N} T_{ji} e_j \\
&= \sum_{j=1}^{N} [\sum_{i=1}^{N} T_{ji} x_i] e_j.
\end{aligned} \tag{A.9}$$

Thus, once we chose our basis then to every linear transformation we assign an $N \times N$ array of numbers which we will call a matrix:

$$\hat{T} \quad \leftrightarrow \quad \boldsymbol{T}$$

$$= \begin{pmatrix} T_{11} & T_{12} & \cdots & T_{1N} \\ T_{21} & T_{22} & \cdots & T_{2N} \\ \vdots & \vdots & \cdots & \vdots \\ T_{N1} & T_{N2} & \cdots & T_{NN} \end{pmatrix}. \tag{A.10}$$

Definition A.12 If $\boldsymbol{T}$ is a $N \times M$ matrix with components $T_{ij}, 1 \le i \le N, 1 \le j \le M$ and $\boldsymbol{B}$ is a $M \times R$ matrix with components B_{qp} with $1 \le q \le M, 1 \le p \le R$ then the matrix $\boldsymbol{A}$ with components $A_{st} = \sum_{k=1}^{M} T_{sk} R_{kt}$, where $1 \le s \le N, 1 \le t \le R$ is a $N \times R$ matrix known as the product matrix

$$\boldsymbol{A} = \boldsymbol{T} \boldsymbol{R}.$$

Notice that for a matrix $\boldsymbol{T}$ the element T_{ij} corresponds to the jth column and ith row. Just as in $\mathbb{R}^3$ we can write $\boldsymbol{r} \in \mathbb{C}^N$ as an ordered N tuple

$$\boldsymbol{r} = (x_1, x_2, \ldots x_N). \tag{A.11}$$

It is now, however, expedient to write it as a column vector, i.e., a $N \times 1$ matrix rather than a row vector, a $1 \times N$ matrix. With this identification (A.9) can be written

$$\hat{T}(\boldsymbol{r}) = \boldsymbol{T} \boldsymbol{r}. \tag{A.12}$$

In summary, for a N-dimensional vector space with a fixed orthonormal basis then

- to each vector there is a one to one correspondence to an N-tuple;

- to each linear operator there is a one to one correspondence with a $N \times N$ matrix;

- the vector $\hat{T}[\boldsymbol{r}]$ corresponding to the image of $\boldsymbol{r}$ under the linear transformation $\hat{T}$ corresponds to N-tuple got by multiplying the $\boldsymbol{r}$ N-tuple by the matrix representation of the operator;

- the inner product corresponds to the multiplication of $1 \times N$ matrix by a $N \times 1$ matrix.

TRANSFORMATIONS FROM $\mathbb{R}^N \mapsto \mathbb{R}^M$

While I will reserve that the term "*linear operator*" only for a linear transformations from a vector space onto itself, we could have a transformation $\hat{A}$ from $\mathbb{R}^N \mapsto \mathbb{R}^M$ with associated matrix $\boldsymbol{A}$ having M rows and N columns, i.e., it is a $M \times N$ matrix. This matrix acts on a vector in $\mathbb{R}^N$, a $N \times 1$ matrix, and converts it into a $M \times 1$ matrix: a vector in $\mathbb{R}^M$, the set of all such vectors is called the "*image of $\hat{A}$*," which I will denote by Ω_A.

Now Ω_A is a subspace of $\mathbb{R}^M$ whose dimension is R Let $\{e_i\}_{i=1}^N$ be the standard basis for $\mathbb{R}^N$. Now if $x, \ \epsilon \ \Omega_A$ then there exists $c \ \epsilon \ \mathbb{R}^N$ s.t.

$$
\begin{aligned}
Ac &= x, \\
c &= \sum_{i=1}^n \alpha_i e_i, \\
\Rightarrow Ac &= \sum_{i=1}^n \alpha_i A e_i, \\
\Rightarrow x &= \sum_{i=1}^n \alpha_i A e_i.
\end{aligned}
$$

Therefore, the set $\{Ae_i\}_{i=1}^N$ must span Ω_A. Further the vector Ae_i is the ith column of A. Hence,

$$R = \dim \left(\mathrm{Span}\{Ae_i\}_{i=1}^N \right),$$

and R is the number number of linearly independent columns of A.

Definition A.13   The column rank of A is the maximal number of linearly independent columns of A. The row rank of A is the maximal number of linearly independent rows of A.

Theorem A.14   *The row rank of a matrix A is equal to its column rank [32].*

Let $\hat{I}$ be the linear operator acting on the finite dimensional vector space $\mathbb{V}$ defined by:

$$
\begin{aligned}
\hat{I} : \mathbb{V} &\mapsto \mathbb{V}, \\
\hat{I}[r] &= r, \text{ for all } r \ \epsilon \ \mathbb{V}.
\end{aligned}
\tag{A.13}
$$

Then from (A.8) then the elements of the matrix representation of $\hat{I}$ are given by

$$I_{ij} = \langle e_i | \hat{I} e_j \rangle = \langle e_i | e_j \rangle = \delta_{ij}, \tag{A.14}$$

i.e.,

$$
I = \begin{pmatrix}
1 & 0 & \cdots & 0 \\
0 & 1 & \cdots & 0 \\
\vdots & \vdots & \cdots & \vdots \\
0 & 0 & \cdots & 1
\end{pmatrix}.
\tag{A.15}
$$

Lemma A.15   *For any square matrix* $\boldsymbol{B}$

$$\boldsymbol{BI} = \boldsymbol{IB} = \boldsymbol{B}.$$

Proof. Let $\boldsymbol{T} = \boldsymbol{BI}$. Then,

$$T_{ij} = \sum_{q=1}^{N} B_{iq}\delta_{qj} = B_{ij} = \sum_{p=1}^{N} \delta_{ip} B_{pj}.$$

$\square$

Definition A.16   An $N \times N$ matrix $\boldsymbol{B}$ has an inverse, $\boldsymbol{B}^{-1}$, if

$$\boldsymbol{BB}^{-1} = \boldsymbol{I} = \boldsymbol{B}^{-1}\boldsymbol{B}.$$

It is trivial to see that $\boldsymbol{B}^{-1}$ is unique.

Before we can proceed and actually construct an inverse matrix we need to take a quick tour through the theory of linear equations. Let us begin with the simplest case suppose we want to solve the set of two simultaneous equations in x and y

$$a_1 x + b_1 y = c_1, \tag{A.16}$$

$$a_2 x + b_2 y = c_2. \tag{A.17}$$

(A.16) and (A.17) are clearly equivalent to the matrix equation

$$\overset{\boldsymbol{Tr} \;=\; c}{\begin{pmatrix} a_1 & b_1 \\ a_2 & b_2 \end{pmatrix} \begin{pmatrix} x \\ y \end{pmatrix} = \begin{pmatrix} c_1 \\ c_2 \end{pmatrix}}, \tag{A.18}$$

and clearly we can solve the set of linear equations iff $\boldsymbol{T}^{-1}$ exists. If we multiply (A.16) by a_2 and (A.17) by a_2 and then subtract we find

$$\begin{aligned} x &= \frac{b_2 c_1 - b_1 c_2}{\det|\boldsymbol{T}|}, \\ y &= \frac{a_1 c_2 - c_1 a_2}{\det|\boldsymbol{T}|}, \end{aligned} \tag{A.19}$$

where we have introduced the determinant of $\boldsymbol{T}$ which is given by

$$\det|\boldsymbol{T}| = a_1 b_2 - a_2 b_1. \tag{A.20}$$

If $\det|\boldsymbol{T}| = 0$ then we are in trouble but if it is non zero then we have solved the set of linear equations. if $\det|\boldsymbol{T}| = 0$ and

$$b_2 c_1 - b_1 c_2 = 0$$
$$\text{and}$$
$$a_1 c_2 - c_1 a_2 = 0,$$

then there is some hope but in this case (A.16) and (A.17) are essentially the same equation and we have only one equation of two unknowns and thus an infinity of solutions. The system of linear equations (A.18) has a unique solution iff $\det[\boldsymbol{T}] \neq 0$. These results can be generalized, a determinant can be defined for $N \times N$ matrix as follows.

Definition A.17   The determinant of the $N \times N$ matrix is defined inductively as follows: it is a linear combination of products of the elements of any row (or column)and the $N - 1$ determinant formed by striking out the row and column of the original determinant in which the element appeared. The reduced array is called a minor and the sign associated with this product is $(-)^{i+j}$. The product of the minor with this sign is called the cofactor. We can keep doing this until we get down to a sum of 2×2 determinants which can be evaluated using (A.20).

Example A.18

$$\begin{vmatrix} a_{11} & a_{12} & a_{13} \\ a_{21} & a_{22} & a_{23} \\ a_{31} & a_{32} & a_{33} \end{vmatrix} =$$

$$(-1)^{1+1} a_{11} \begin{vmatrix} a_{22} & a_{23} \\ a_{32} & a_{33} \end{vmatrix} + (-1)^{12} a_{12} \begin{vmatrix} a_{21} & a_{23} \\ a_{31} & a_{33} \end{vmatrix} + (-1)^{13} a_{13} \begin{vmatrix} a_{21} & a_{22} \\ a_{31} & a_{32} \end{vmatrix}$$

$$= a_{11}[a_{22}a_{33} - a_{23}a_{32}] - a_{12}[a_{31}a_{33} - a_{23}a_{31}] + a_{13}[a_{21}a_{32} - a_{22}a_{31}]. \tag{A.21}$$

It can be shown [32] that if $\boldsymbol{A}$ is a $N \times N$ matrix then it has a unique inverse iff $\det[\boldsymbol{A}] \neq 0$ iff $\text{rank}(\boldsymbol{A}) = N$, this last condition is equivalent to saying that its columns(rows) treated as vectors must be linearly independent. In fact, it can be shown that

$$\boldsymbol{B}^{-1} = \frac{\boldsymbol{C}^T}{|\boldsymbol{B}|}, \tag{A.22}$$

where C is the cofactor matrix, constructed as follows. The ij element of the cofactor matrix C is c_{ij} which is $(-1)^{i+j}$ multiplied by determinants of the $(N-1) \times (N-1)$ matrix got by striking out the ith row and jth column of the original matrix B. C^T denotes the transpose of C.

Definition A.19 Let $\hat{T}$ be an operator defined on a vector space, $\mathbb{V}$, upon which an inner product is defined. We define the adjoint of $\hat{T}$ to be a linear operator $\hat{T}^\dagger : \mathbb{V} \mapsto \mathbb{V}$ where for all $a, b \in \mathbb{V}$

$$\langle a | \hat{T} b \rangle = \langle \hat{T}^\dagger a | b \rangle.$$

Lemma A.20 *If $\hat{T}$ is a linear operator acting on an N dimensional vector space, $\mathbb{V}$ with matrix representation, $T \equiv (T)_{ij}$ then its adjoint $T^\dagger$ has the matrix representation $\bar{T}_{ji}$, i.e., we interchange rows and columns and take the complex conjugate of each element.*

Proof.

$$
\begin{aligned}
(T)_{ij} &= \langle e_i | \hat{T} e_j \rangle \\
&= \langle \hat{T}^\dagger e_i | e_j \rangle \\
\\
&= \overline{\langle e_j | \hat{T}^\dagger e_i \rangle} \\
\Rightarrow (T^\dagger)_{ij} &= (\bar{T})_{ji}.
\end{aligned}
$$

$\square$

To be clear, if we start we an operator $\hat{T}$ with a matrix representation given by (A.10) then its adjoint $\hat{T}^\dagger$ has a matrix representation given by

$$
\hat{T}^\dagger \quad \leftrightarrow \quad T^\dagger
$$
$$
= \begin{pmatrix}
\bar{T}_{11} & \bar{T}_{21} & \cdots & \bar{T}_{N1} \\
\bar{T}_{12} & \bar{T}_{22} & \cdots & \bar{T}_{N2} \\
\vdots & \vdots & \cdots & \vdots \\
\bar{T}_{1N} & \bar{T}_{2N} & \cdots & \bar{T}_{NN}
\end{pmatrix}. \tag{A.23}
$$

Lemma A.21 *Let A be an $M \times M$ complex matrix and B be a $M \times M$ complex matrix then*

$$(AB)^\dagger = B^\dagger A^\dagger.$$

Proof. Looking at components,

$$
\begin{aligned}
(\boldsymbol{AB})_{ij} &= \sum_{q=1}^{M} a_{iq}b_{qj}, \\
(\boldsymbol{AB})_{ij}^{\dagger} &= \overline{(\boldsymbol{AB})_{ji}} \\
&= \sum_{q=1}^{M} \bar{a}_{jq}\bar{b}_{qi} \\
&= \sum_{q=1}^{M} (\boldsymbol{B}^{\dagger})_{iq}(\boldsymbol{A}^{\dagger})_{qj} = (\boldsymbol{B}^{\dagger}\boldsymbol{A}^{\dagger})_{ij}.
\end{aligned}
$$

$\square$

Definition A.22   Let $\hat{T}$ be an operator defined on a vector space, $\mathbb{V} \mapsto \mathbb{V}$ if $\hat{T} = \hat{T}^{\dagger}$, then the operator is said to be self-adjoint. We know that the matrix representation of the adjoint matrix is given by

$$
T_{ij}^{\dagger} = \bar{T}_{ji}.
$$

If $\hat{T}$ is self-adjoint then it follows that

$$
T_{ij} = \bar{T}_{ji}, \tag{A.24}
$$

such a matrix is said to be hermetian.

Definition A.23   A non-zero vector $\boldsymbol{a}$ is an eigenvector of $\hat{T}$ with eigenvalue λ if the effect of acting with the operator is simply to multiply the vector by λ, i.e.,

$$
\hat{T}\boldsymbol{a} = \lambda\boldsymbol{a}.
$$

Lemma A.24   *If $\hat{T}$ is a self-adjoint operator defined $\mathbb{V} \mapsto \mathbb{V}$, then its eigenvalues must be real.*

Proof. Let $\boldsymbol{a}$ be an eigenvector of $\hat{T}$ with eigenvalue λ. Note we have excluded the null vector from being an eigenvector but we have not excluded the number zero for being an eigenvalue. Consider

$$
\begin{aligned}
\langle \boldsymbol{a}|\hat{T}\boldsymbol{a}\rangle &= \langle \boldsymbol{a}|\lambda\boldsymbol{a}\rangle = \lambda\langle \boldsymbol{a}|\boldsymbol{a}\rangle = \lambda\|\boldsymbol{a}\|^{2}, \\
\langle \hat{T}^{\dagger}\boldsymbol{a}|\boldsymbol{a}\rangle &= \langle \hat{T}\boldsymbol{a}|\boldsymbol{a}\rangle = \langle \lambda\boldsymbol{a}|\boldsymbol{a}\rangle = \bar{\lambda}\|\boldsymbol{a}\|^{2}, \\
\Rightarrow \lambda &= \bar{\lambda}.
\end{aligned}
$$

$\square$

Lemma A.25   *If $\{b_i\}_{i=1}^{M}$ are the eigenvectors of a self adjoint operator $\hat{B}$ corresponding to distinct eigenvalues $\{\beta_i\}_{i=1}^{M}$, then these eigenvectors are orthogonal.*

Proof. Suppose b_i, b_j are eigenvectors corresponding to eigenvalues $\beta_1, \beta_j, \beta_i \neq \beta_j$. Remembering that the eigenvalues are real, we can write

$$
\begin{aligned}
\langle b_i | \hat{B} b_j \rangle &= \beta_j \langle b_i | b_j \rangle \\
= \langle \hat{B} b_i | b_j \rangle &= \beta_i \langle b_i | b_j \rangle, \\
\Rightarrow [\beta_i - \beta_j] \langle b_i | b_j \rangle &= 0, \\
\Rightarrow \langle b_i | b_j \rangle &= 0.
\end{aligned}
$$

$\square$

These two lemmas though simple to prove turn out to be very important. We have proved the results for a general operator rather than just the matrix representation, so they will hold in any finite or infinite dimensional vector space. Suppose we are working in an N-dimensional space.

Theorem A.26   *If B is an $N \times N$ matrix its eigenvalues are the solution of the equation*

$$
\det[B - \beta I] = 0.
$$

Proof. Suppose b is an eigenvector, with eigenvalue β, then

$$
\begin{aligned}
Bb &= \beta b, \\
\Rightarrow [B - \beta I] b &= 0.
\end{aligned}
$$

If $[B - \beta I]^{-1}$ exists then if we act with it we find that $b = 0$, i.e., no eigenvectors exist so for us to find eigenvalues we must have

$$
\det[B - \beta I] = 0.
$$

This will yield a polynomial of order N in β and by the fundamental theorem of algebra this has N complex roots which is the maximal number of eigenvalues possible. $\square$

Suppose B is a self-adjoint operator with a maximal set of distinct eigenvalues $\{\beta_i\}_{i=1}^{N}$ with associated eigenvectors $\{b_i\}_{i=1}^{N}$. The eigenvectors may be written

$$
b_i = \begin{pmatrix} b_{1i} \\ \vdots \\ b_{Ni} \end{pmatrix},
$$

$$\sum_{k=1}^{N} \bar{b}_{ki} b_{kj} = \delta_{ij}. \tag{A.25}$$

If we define a matrix S whose columns are the eigenvectors of B, i.e.,

$$S_{ij} \equiv b_{ij}, \tag{A.26}$$

then (A.25) is equivalent to:

$$S^{-1} = S^{\dagger}. \tag{A.27}$$

An operator which satisfies the property (A.27) is said to be unitary. I will have a lot more to say about such operators below.

Consider:

$$
\begin{aligned}
[S^{\dagger} B S]_{ij} &= \sum_{k=1}^{N} S_{ik}^{\dagger} (\sum_{p=1}^{N} B_{kp} S_{pj}) \\
&= \sum_{k=1}^{N} \bar{b}_{ki} (\sum_{p=1}^{N} B_{kp} b_{pj}) \\
&= \sum_{k=1}^{N} \bar{b}_{ki} \lambda_j b_{kj}, \\
&= \lambda_j \delta_{ij}, \tag{A.28}
\end{aligned}
$$

or in matrix form

$$S^{\dagger} B S = \begin{pmatrix} \lambda_1 & 0 & 0 & \cdots & 0 \\ 0 & \lambda_2 & 0 & \cdots & 0 \\ \vdots & \vdots & \vdots & \ddots & \vdots \\ 0 & 0 & 0 & \cdots & \lambda_N \end{pmatrix}. \tag{A.29}$$

Our derivation of (A.29) depended on the eigenvectors being mutually orthogonal, the proof of which depended on the eigenvalues being distinct. It is not unusual to find a self-adjoint operator B which has more than one eigenvectors which are linearly independent of each other but have the same eigenvalue. Suppose the operator $\hat{B}$ has M eigenvectors $\{b_i\}_1^M$ such that each of them satisfies

$$\hat{B} b_i = \beta b_1. \tag{A.30}$$

Consider

$$a = \sum_{i=1}^{m} \alpha_i b_i,$$

where α_i are complex numbers then

$$
\begin{aligned}
\hat{B}[a] &= \sum_{i=1}^{M} \alpha_i \hat{B}[b_i] \\
&= \sum_{i=1}^{M} \alpha_i \beta[b_i] \\
&= \beta a.
\end{aligned} \tag{A.31}
$$

Thus, the set $\Omega = \{$eigenvectors of $\hat{B}$ with eigenvalue $\beta\}$ is itself a vector space which is a subspace of our original space. We may chose a maximal set of M, say, linearly independent vectors which we can orthognalize to each other and to the other eigenfunctions of $\hat{B}$ using our Grahm–Schmidt processes. We can repeat this processes for any other degenerate eigenvalues until we arrive at a maximal set of mutually orthogonal eigenvectors.

CHANGE OF BASIS

The matrix we have constructed in (A.27) we described as unitary.

Definition A.27   A linear operator $\hat{U}$ is said to be unitary if

$$
\hat{U}^\dagger = \hat{U}^{-1}.
$$

Lemma A.28   *If U is a linear transformation then the following are equivalent.*

(a) U is unitary.

(b) For every $x \in \mathbb{V}$

$$
\|Ux\| = \|x\|.
$$

(c) For every $x, y \in \mathbb{V}$

$$
\langle Ux | Uy \rangle = \langle x | y \rangle.
$$

Proof. In this proof we will make use of the following identity:

$$
\begin{aligned}
\frac{1}{4}\left[\langle x + y | x + y \rangle - \langle x - y | x - y \rangle \right] &= \frac{1}{4}\left[\|x\|^2 + \|y\|^2 + 2\langle x|y\rangle - \right. \\
&\qquad \left. \|x\|^2 - \|y\|^2 + 2\langle x|y\rangle \right] \\
&= \langle x | y \rangle.
\end{aligned} \tag{A.32}
$$

If U is unitary then (c) follows since

$$\langle Ux|Uy\rangle = \langle U^{-1}Ux|y\rangle = \langle x|y\rangle.$$

and (b) is a special case of (c) so we have $(a) \Rightarrow (b) \Rightarrow (c)$.

Suppose (b) holds. Consider:

$$
\begin{aligned}
\langle Ux|Uy\rangle &= \frac{1}{4}\left[\langle Ux + Uy|Ux + Uy\rangle - \langle Ux - Uy|Ux - Uy\rangle\right] \\
&= \frac{1}{4}\left[\|U(x + y)\|^2 - \|U(x - y)\|^2\right] \\
&= \frac{1}{4}\left[\|x + y\|^2 - \|x - y\|^2\right] \\
&= \langle x|y\rangle.
\end{aligned}
$$

So $(b) \Rightarrow (c)$.

Suppose (c) then

$$
\begin{aligned}
\langle Ux|Uy\rangle &= \langle x|y\rangle \\
&= \langle U^\dagger Ux|y\rangle, \\
\Rightarrow \langle x - U^\dagger Ux|y\rangle &= 0, \ \forall \, y, \\
\Rightarrow x &= U^\dagger Ux, \ \forall \, x, \\
\Rightarrow U^\dagger U &= I, \\
\Rightarrow (c) &\Rightarrow (a).
\end{aligned}
$$

$\square$

Lemma A.29   *Suppose A is an $N \times N$ matrix and U is a unitary matrix then if $B = UAU^\dagger$ then B and A have the same eigenvalues.*

Proof. Suppose r is an eigenvector of A with eigenvalue λ

$$
\begin{aligned}
Ar &= \lambda r, \\
\Rightarrow U^\dagger BUr &= \lambda r, \\
\Rightarrow BUr &= \lambda Ur.
\end{aligned}
$$

Thus, λ is an eigenvalue of B corresponding to the eigenvector Ur. $\square$

APPENDIX B

Analytic Solution to the Quantum Oscillator

The harmonic oscillator is one of the few quantum systems that admits a relatively simple analytic solution [26]. We use this solution to benchmark our numerical code. We start with the Schrödinger equation

$$-\frac{\hbar^2}{2m}\frac{d^2\psi(x)}{dx^2} + \frac{1}{2}m\omega^2 x^2\psi(x) = E\psi(x). \tag{B.1}$$

We will work with units where $\hbar = 1, m = 1, \omega = 1$. In these units,

$$\frac{d^2\psi(x)}{dx^2} = (x^2 - 2E)\psi. \tag{B.2}$$

This formal differential equation, (B.2), needs to augmented by boundary conditions. In order to retain the probability interpretation [23], we must require that the function $\psi(x)$ be square integrable, i.e., we must have

$$\int_{-\infty}^{\infty} |\psi(x)|^2 dx$$

be finite. A necessary condition is that

$$\lim_{x\to\pm\infty} \psi(x) \to 0. \tag{B.3}$$

Equation (B.3) is a second-order differential equation and as such will admit two linearly independent solutions. Only one of which will be consistent with the boundary condition (B.3). Asymptotically, for $x >> 1$ we can approximate

$$\frac{d^2\psi(x)}{dx^2} \approx x^2\psi(x), \tag{B.4}$$

which has solutions

$$\psi_\pm(x) = e^{\pm x^2/2}. \tag{B.5}$$

Notice

$$\lim_{x\to\infty} \psi_-(x) \to 0,$$

$$\lim_{x \to \infty} \psi_+(x) \ \to \ \infty. \tag{B.6}$$

Clearly, we don't want the divergent solution $\psi_+(x)$. Returning to the full differential equation, (B.1), let us look to see if we can find a solution of the form:

$$\psi(x) = h(x)e^{-x^2/2}, \tag{B.7}$$

where $h(x)$ is some analytic function with a power series expansion

$$h(x) = \sum_{j=0}^{\infty} a_j x^j, \tag{B.8}$$

from which it follows

$$
\begin{aligned}
h'(x) &= \sum_{j=0}^{\infty} a_j j x^{j-1}, \\
h''(x) &= \sum_{j=0}^{\infty} a_j j(j-1)x^{j-2}, \\
\Rightarrow \frac{d\psi(x)}{dx} &= \frac{dh(x)e^{-x^2/2}}{dx} \\
&= h'(x)e^{-x^2/2} - xh(x)e^{-x^2/2}, \\
\frac{d^2\psi(x)}{dx^2} &= h''(x)e^{-x^2/2} - xh'(x)e^{-x^2/2} - h(x)e^{-x^2/2} - x(h'(x)e^{-x^2/2} - xh(x)e^{-x^2/2}) \\
&= h''(x)e^{-x^2/2} - 2xh'(x)e^{-x^2/2} - h(x)e^{-x^2/2} + x^2 h'(x)e^{-x^2/2} \\
&= e^{-x^2/2}\left[\sum_{j=0}^{\infty} a_j j(j-1)x^{j-2} - 2\sum_{j=0}^{\infty} a_j j x^j - \sum_{j=0}^{\infty} a_j x^j + \sum_{j=0}^{\infty} a_j j x^{j+1}\right] \\
&= e^{-x^2/2}\left[\sum_{j=0}^{\infty}(a_{j+2}(j+2)(j+1) - 2a_j)x^j\right]. \tag{B.9}
\end{aligned}
$$

Thus,

$$
\frac{d^2\psi(x)}{dx^2} - x^2\psi(x) + 2E\psi(x) = 0,
$$

$$
\Rightarrow \left[\sum_{j=0}^{\infty}(a_{j+2}(j+2)(j+1)) - 2a_j j + (2E-1)a_j)x^j\right] = 0. \tag{B.10}
$$

Now as we noted earlier power series coefficients are unique so

$$
(a_{j+2}(j+2)(j+1)) - 2a_j j \ + \ (2E-1)a_j) = 0,
$$

$$\Rightarrow a_{j+2} = \frac{(2j + 1 - 2E)a_j}{(j + 1)(j + 2)}. \tag{B.11}$$

Thus, if we know a_0 we can find all even coefficients and have a_1 we can find all odd coefficients. We may write

$$h(x) = h_{even}(x) + h_{odd}(x). \tag{B.12}$$

Since we are interested in asymptotics we can concentrate on the larger powers, $j \gg 1$ for which

$$a_{j+2} \approx \frac{2}{j}a_j,$$

thus

$$\begin{aligned} h(x) &= h_{even}(x) + h_{odd}(x) \\ &= a_0(1 + \sum_{j=1}^{\infty} \frac{2}{j!}x^{2j}) + a_1(x + \sum_{j=1}^{\infty} \frac{2}{(j+1)!}x^{2j+1}). \end{aligned} \tag{B.13}$$

This diverges like e^{+x^2} the solution that we didn't want. But if there exists an integer j such that

$$2j + 1 = 2E \tag{B.14}$$

then one of the series will terminate and we can set either a_0 or a_1 equal to zero to get rid of the diverging series. The resulting finite series solution will has the correct asymptotic form. The normalized eigenfunctions are

$$\phi_n(x) = \frac{1}{\sqrt{2^n n!}} H_n(x) e^{-\frac{x^2}{2}}, \tag{B.15}$$

where $H_n(x)$ is a hermite polynomial. The first few are given by

$$\begin{aligned} H_0(x) &= 1, \\ H_1(x) &= 2x, \\ H_2(x) &= 4x^2 - 2, \\ H_3(x) &= 8x^3 - 12x. \end{aligned} \tag{B.16}$$

Notice that $H_0(x)$, $H_2(x)$ are even functions of x while $H_1(x)$, $H_3(x)$ are odd. In Figure B.1 the first four eigenfunctions are plotted.

As expected, the lowest eigenfunction, corresponding to E_0 is even with no zeros, ϕ_1 is odd with one zero, ϕ_2 even with two zeros, and ϕ_3 is odd with three zeros.

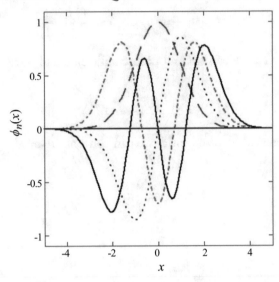

Figure B.1: The first harmonic oscillator eigenfunctions for unit frequency, $\omega = 1$, in units where $\hbar = m = 1$: $n = 0$, dashed red, $n = 1$ dotted blue, $n = 2$, dashed dotted green, $n = 3$ solid black.

Equation (B.14) leads us to the allowed energy eigenvalues

$$E \;=\; j + \frac{1}{2}. \tag{B.17}$$

If j is odd we must take $a_0 = 0$ and if j is even we must take $a_1 = 0$ and we recover the Hermite polynomials which as you will recall contain only odd or even powers of x, thus the associated eigenfunctions, $\phi_n(x)$, are such that

$$\begin{aligned}
\phi_{2n}(-x) &\;=\; \phi_{2n}(x), \\
\phi_{2n+1}(-x) &\;=\; -\phi_{2n+1}(x).
\end{aligned} \tag{B.18}$$

APPENDIX C

First-Order Perturbation Theory

Suppose we have a Hamiltonian $\hat{H}_0$ which has a known set of eigenvalues, E_j, with an associated set of orthonnormal eigenvectors $\{\psi_j^0\}$. We will assume there is no degeneracy, i.e., $E_i \neq E_j$ if $i \neq j$. If ψ is any state of the system then

$$\psi = \sum_j a_j \psi_j^0. \tag{C.1}$$

Now, suppose our system is *"perturbed"* by a small extra potential so we will have another Hamiltonian $\hat{H}$ which is *"not too different"* from $\hat{H}_0$. We can write

$$\hat{H} = \hat{H}_0 + \hat{H}_I,$$

where $\hat{H}_I = \lambda \hat{H}_1$ where λ is "quadratically small;" in other words, λ^2 is negligibly small. For example, if we place a one electron atom in an electric field, $\boldsymbol{E} = \mathbb{E}\boldsymbol{e}_z$ the Hamiltonians are

$$\begin{aligned} \hat{H}_0 &= \frac{\hbar^2}{2\mu}\nabla^2 - \frac{Z}{r}, \\ \hat{H}_1 &= -z. \end{aligned} \tag{C.2}$$

We could reasonably assume that if $\lambda = \mathbb{E}$ is small the effect on the energy levels will also be small, i.e., we would expect that the eigenenergies of the new Hamiltonian would be very similar to the original; in other words, if $\bar{E}_i$ is a new eigenvalue then:

$$\begin{aligned} \hat{H}\psi_i &= \bar{E}_i\psi, \\ \bar{E}_i &= E_i + \Delta E_i, \\ |\Delta E_i| &<< 1. \end{aligned} \tag{C.3}$$

We also expect that the new eigenvector, ψ_i, will be *"not too different"* from the original ψ_i^0. To be a little more precise, we can expand:

$$\psi_i = \sum_j c_{ij}\psi_j^0, \tag{C.4}$$

and since we require $\langle \psi_i | \psi_i \rangle = 1$ we have

$$\sum_j |c_{ij}|^2 = 1. \tag{C.5}$$

We require

$$\begin{aligned} c_{ii} &\approx 1, \\ c_{ij} &\sim 0 \quad i \neq j, \end{aligned} \tag{C.6}$$

i.e., we require c_{ii} to differ from 1 by a quadratically small quantity. The eigenvalue equation for ψ_i is

$$\begin{aligned} H\psi_i &= \bar{E}_i \psi_i, \\ \left[\hat{H}_0 + \lambda \hat{H}_1 \right] \psi_i &= (E_i + \Delta E_i)\psi_i, \\ \left[\hat{H}_0 + \lambda \hat{H}_1 \right] \sum_j c_{ij} \psi_j^0 &= (E_i + \Delta E_i) \sum_j c_{ij} \psi_j^0, \\ \sum_j c_{ij}(E_j \psi_j^0 + \lambda \hat{H}_1 \psi_j^0) &= E_i \sum_j c_{ij} \psi_j^0 + \Delta E_i \sum_j c_{ij} \psi_j^0, \\ \sum_{j \neq i} c_{ij}[E_j - E_i]\psi_j^0 + \sum_j \lambda \hat{H}_1 c_{ij} \psi_j^0 &= \sum_j \Delta E_i c_{ij} \psi_j^0. \end{aligned} \tag{C.7}$$

Let us now neglect quadratically small quantities λc_{ij} and $\Delta E c_{ij}$ when $i \neq j$ and assume $c_{ii} \approx 1$ then taking the inner product with ψ_i^0 we have:

$$\langle \psi_i^0 | \hat{H}_I \psi_i^0 \rangle \approx \Delta E_i. \tag{C.8}$$

That is the shift ΔE_i in the level E_i resulting from the addition of the perturbation $\hat{H}_I$ to the original Hamiltonian is just the expectation value of the perturbing Hamiltonian calculated from the original eigenket ψ_i^0. For our purposes, here we will not need more than the energy shift ΔE_i, however if we take an inner product with ψ_i^0 on (C.7) we find that

$$c_{iq} = \frac{\langle \psi_q^0 | \hat{H}_I \psi_i^0 \rangle}{E_i - E_q} \quad i \neq q, \tag{C.9}$$

which together with $c_{ii} = 1$ gives us an estimate for ψ_i. Notice our original assumption of non-degeneracy means that (C.9) is well defined. If level i is degenerate then our assumption that all c_{ij} $i \neq j$ are quadratically small may not hold. We do not need all the eigenvalues to be non-degenerate only the particular E_i we want to study. The approximation (C.8) is correct only to first order in small quantities.

Bibliography

[1] William H. Press, Saul A. Teukolsky, William T. Vetterling, and Brian P. Flannery. *Numerical Recipes, Fortran:77*. Cambridge University Press, Cambridge, 1992. 4

[2] LAPACK (linear algebra package) is a standard software library for numerical linear algebra. www.netlib.org/lapack 52

[3] The Numerical Algorithm Group (NAG) library is a commercial software library. https://www.nag.co.uk/content/nag-library-fortran 4

[4] CASTEP is a shared source suite for calculating the electronic properties of crystalline solids, surfaces, molecules, liquids and amorphous materials from first principles. www.castep.org 4

[5] Quantum ESPRESSO is a suite for first-principles electronic-structure calculations and materials modeling. https://www.quantum-espresso.org

[6] Gaussian is a general purpose computational chemistry software package. www.gaussian.com

[7] K. G. Dyall, I. P. Grant, C. T. Johnson, F. A Parpia, and E. P. Plummer. GRASP: A general-purpose relativistic atomic structure program. *Comput. Phys. Commun.*, 55:425, 1989. DOI: 10.1016/0010-4655(89)90136-7

[8] R. J. Needs, M. D. Towler, N. D. Drummond, and P. Lopez Rios. Continuum variational and diffusion quantum Monte Carlo calculations. *J. Phys. Condens. Matter*, 22:023201, 2010. DOI: 10.1088/0953-8984/22/2/023201 4

[9] Dennis M. Ritchie and Brian W. Kernighan. *The C Programming Language*, 2nd ed., Prentice Hall, 1988. DOI: 10.1007/978-3-662-09507-2_22 4

[10] John V. Guttag. *Introduction to Computation and Programming Using Python: With Application to Understanding Data*. MIT Press, 2016. 4

[11] Colm T. Whelan. *A First Course in Mathematical Physics*. Wiley-VCH, 2016. 5, 21, 33, 34, 63, 92, 97, 123

[12] S. Lang. *Analysis I*. Addison-Wesley, New York, 1968. 6, 11, 21

[13] Hans J. Weber and George B. Arfken. *Essential Mathematical Methods for Physicists*. Elsevier, 2004. 11, 66, 67, 71

[14] Milton Abramowitz and Irene A. Stegun. *Handbook of Mathematical Functions with Formulas, Graphs, and Mathematical Tables*. Dover/National Bureau of Standard, 1972. DOI: 10.1115/1.3625776 16

[15] Endre Süli and David Mayers. *An Introduction to Numerical Analysis*. Cambridge University Press, 2003. DOI: 10.1017/cbo9780511801181 28, 47, 53, 55, 63

[16] Donald L. Kreider, Robert G. Kuller, Donald R. Ostberg, and Fred W. Perkins. *An Introduction to Linear Analysis*. Addison-Wesley, Reading, 1966. DOI: 10.2307/2313834 33, 38

[17] John R. Taylor. *Classical Mechanics*. University Science Books, 2005. 33, 44, 97, 98

[18] J. H. Wilkinson. *The Algebraic Eigenvalue Problem*. Oxford, 1965. DOI: 10.1007/978-93-86279-52-1_11 52

[19] James W. Longley. Modified Gram-Schmidt process vs. classical Gram-Schmidt. *Communic. Statist. Simul. Computat.*, 10(5):517, 1981. DOI: 10.1080/03610918108812227 58

[20] Colm T. Whelan. On the Bethe approximation to the reactance matrix. *J. Phys. B*, 19:2343, 1986. DOI: 10.1088/0022-3700/19/15/015 67

[21] Alan Burgess and Colm T. Whelan. BETRT—a procedure to evaluate cross-sections for electron hydrogen collisions in the Bethe approximation to the reactance matrix. *Comput. Phys. Commun.*, 47:295, 1987. DOI: 10.1016/0010-4655(87)90115-9 67

[22] Earl A. Coddington and Norman Levinson. *Ordinary Differential Equations*. MGraw-Hill, 1984. DOI: 10.1063/1.3059875 79

[23] Colm T. Whelan. *Atomic Structure*. IOP-Morgan & Claypool, 2018. DOI: 10.1088/978-1-6817-4880-1 83, 92, 107, 108, 109, 110, 115, 135

[24] C. Froese-Fischer. *The Hartree-Fock Method for Atoms—a Numerical Approach*. Wiley, 1977. 115

[25] Reiner M. Dreizler and Eberhard K. U. Gross. *Density Functional Theory*. Springer, 1999. DOI: 10.1007/978-3-642-86105-5 92

[26] David J. Griffiths. *Introduction to Quantum Mechanics*, 2nd ed., Cambridge University Press, 2017. DOI: 10.1017/9781316995433 108, 110, 135

[27] A. R. P. Rau. The negative ion of hydrogen. *J. Astrophys Astr.*, 17:113, 1996. DOI: 10.1007/bf02702300 111

[28] Shaun Lucey, Colm T. Whelan, R. J. Allan, and H. R. J. Walters. (e, 2e) on hydrogen minus. *J. Phys. B*, 29(13):L489, 1996. DOI: 10.1088/0953-4075/29/13/002 111

[29] H. Bethe. Berechnung der Elektronenaffinität des Wasserstoffs. *Z. Phys.*, 57:815, 1928. DOI: 10.1007/bf01340659 111

[30] S. Chandrasekar. Some remarks on the negative hydrogen ion and its absorption coefficient. *Astrophys. J.*, 100:176, 1944. DOI: 10.1086/144654 112

[31] Attila Szabo and Neil S. Ostlund. *Modern Quantum Chemistry: Introduction to Advanced Electronic Structure Theory*. Dover, 1996. 112

[32] Serge Lang. *Linear Algebra*. Springer, 1987. DOI: 10.1007/978-1-4757-1949-9 125, 127

Author's Biography

COLM T. WHELAN

Colm T. Whelan is a Professor of Physics and an Eminent Scholar at Old Dominion University in Norfolk, Virginia. He received his Ph.D. in Theoretical Atomic Physics from the University of Cambridge in 1985 and was awarded an Sc.D. also from Cambridge in 2001. He is a Fellow of both the American Physical Society and the Institute of Physics (UK). He has over 30 years of experience in the teaching of physics.

Index

Printed in the United States
by Baker & Taylor Publisher Services